T0250048

PRINCIPLES OF DOWNSTREAM TECHNIQUES IN BIOLOGICAL AND CHEMICAL PROCESSES

PRINCIPLES OF DOWNSTREAM TECHNIQUES IN BIOLOGICAL AND CHEMICAL PROCESSES

Mukesh Doble, PhD, FRSC

AAP APPLE ACADEMIC PRESS

Apple Academic Press Inc. | Apple Academic Press Inc.
3333 Mistwell Crescent | 9 Spinnaker Way
Oakville, ON L6L 0A2 | Waretown, NJ 08758
Canada | USA

©2016 by Apple Academic Press, Inc.

First issued in paperback 2021

Exclusive worldwide distribution by CRC Press, a member of Taylor & Francis Group
No claim to original U.S. Government works

ISBN 13: 978-1-77463-554-4 (pbk)
ISBN 13: 978-1-77188-140-1 (hbk)

Typeset by Accent Premedia Services (www.accentpremedia.com)

Library and Archives Canada Cataloguing in Publication

Doble, Mukesh, author
Principles of downstream techniques in biological and chemical processes / Mukesh Doble, PhD, FRSC.

Includes bibliographical references and index.
Issued in print and electronic formats.
ISBN 978-1-77188-140-1 (hardcover).--ISBN 978-1-4987-3250-5 (pdf)
1. Chemical industry. 2. Chemical engineering. 3. Biochemical engineering.
4. Production management. I. Title.

| TP200.D63 2015 | 660 | C2015-906427-9 | C2015-906428-7 |

Library of Congress Cataloging-in-Publication Data

Doble, Mukesh.
Principles of downstream techniques in biological and chemical processes / Mukesh Doble, PhD, FRSC.

pages cm
Includes bibliographical references and index.
ISBN 978-1-77188-140-1 (alk. paper)
1. Chemical industry. 2. Chemical engineering. 3. Biochemical engineering.
4. Production management. I. Title.

| TP200.D63 2015 | 660--dc23 | 2015035626 |

Apple Academic Press also publishes its books in a variety of electronic formats. Some content that appears in print may not be available in electronic format. For information about Apple Academic Press products, visit our website at **www.appleacademicpress.com** and the CRC Press website at **www.crcpress.com**

CONTENTS

LIST OF ABBREVIATIONS

2D	two-dimensional
ATZ	alumina, titania and zirconia
BSA	bovine serum albumin
CA	cellulose acetate
CSTR	continuous stirred tank reactors
DCF	discounted cash flow
ED	electrodialysis
ERR	economic rate of return
GE	General Electric
GPC	gel permeation chromatography
HETP	height equivalent of a theoretical plate
HIC	hydrophobic-interaction chromatography
HTU	height of transfer units
ICI	imperial chemical industries
IEC	ion exchange chromatography
IEF	isoelectric focusing
IgG	immunoglobulin G
IP	isoelectric point
IRR	internal rate of return
MF	microfiltration
NF	nanofiltration
NPV	net present value
NTU	number of transfer units
PET	polyethylene terephthalate
PSO	polysulphone
PVDF	polyvinylidene difluoride
RO	reverse osmosis
ROI	return on investment
TLC	thin layer chromatography
TPC	total plant cost

TPDC	total plant direct cost
TPIC	total plant indirect cost
UF	ultrafiltration

LIST OF SYMBOLS

$[C]$	amount of soluble protein released as a function of time, t
$[C]_f$	amount of active protein finally recovered
$[C]_o$	maximum amount of soluble protein
$[S_H]$	concentration of solute in the heavy
$[S_L]$	concentration of solute in the light
A	area of contact
A	overall contact area
a	activity
A	cross sectional area of the tube
A	filtration area
A	membrane surface area (cm²)
a	packing area per bed volume
a	surface area of adsorbent per tank volume
A_h and A_c	area on the hot and cold sides, respectively
b_d	bed thickness
b_m	membrane thickness
C	concentration
c	solute concentration
c	total molar concentration
c_0	maximum concentration
C_b	bulk mass concentration of foulant
C_i	molar concentration of species i
C_m	solute concentration in the mobile phase
c_n	solute concentration in the liquid in stage n and also flowing out
C_{n-1}	concentration entering stage n
C_s	solute concentration in the stationary phase
c_s	concentration of the solute that is being filtered near the surface of the membrane
C_T	sum of concentration of all the species
D	agitator diameter

d	degrees of freedom
d	diameter of the spherical particle
D	diffusion coefficient
D	effective diffusion coefficient in the pores
d	particle size or pore diameter in the filter cake
d	wall thickness
D_{AB}	diffusivity of A in B
D_d	dispersion coefficient
D_e	diffusion coefficient of the protein
e	washing efficiency
E	activation energy
E	current efficiency
E_d	activation energy for deactivation of the protein
F	Faraday's constant
F	feed flow rate
F	feed quantity
f	fraction of the bulk concentration that contributes to deposit growth
F1 and F2	streams entering the unit
F3 and F4	streams leaving the unit
fc	crushing strength of the material
Fs	volumetric flow rates
G	crystal growth rate
G	free energy
g_c	driving force
h	heat transfer coefficient
I	current
J_i	flux of component i (moles/h cm²)
k	constant
K	equilibrium constant
k	first order rate constant
K	partitioning coefficient
K	proportionality constant
k	thermal conductivity of the material
K_d	distribution coefficient
K_H	Henry's law constant
K_K	Kick's constant
k_L	gas to liquid mass transfer coefficient

k_L	mass transfer coefficient
K_R	constant
L	length of the packed chromatographic bed
n	constant (if the adsorption is favorable, then n < 1; if it is unfavorable, then n > 1)
n	moles of solute
n	number of cells
N	number of components
n	number of disks
N	number of stages
N	rpm
N	solution normality
n	volume of wash liquid divided by the volume of liquid retained in the cake
N_{Re}	Reynolds number
Nu	Nusselt number
P	operating pressure
P	number of phases
P_0	atmospheric pressure (kg/ cm^2 a)
P_1	initial pressure (kg/ cm^2 a) after bleeding
P_2	final pressure after filling (kg/cm^2 a)
p_A and p_B	partial vapor pressures of component A and B, respectively
$P_{L,i}$	permeability coefficient of component i;
P_o	permeability constant
P_{os}	osmotic pressure
Pr	Prandtl number
Q	rate of heat transfer
q	adsorbed solute concentration
q	amount of solute adsorbed per amount of adsorbent (gm/gm)
Q	flow rate, m^3/s
q	heat transfer per unit time (W)
Q	solvent flow
Q	volumetric flow rate
q and q_F	the final and feed concentrations of the solute in the adsorbent, respectively
Q_i	moles of component i permeated in time t

Q_i and Q_o	blood flow at the inlet and outlet of the dialyzer, respectively
q_n	solute concentration in the adsorbent
Q_o	initial flow rate through the unfouled membrane
q_o and K_e	constants (q_o maximum adsorption capacity and Ke adsorption constant)
r	distance from the axis of rotation
r	distance of the solid particle from the center of the axis of the centrifuge
R	gas constant
r	rate of reaction
R	resistance
R_0 and R_1	the outer and inner radius of the bowl
r_{ads}	rate of adsorption per volume of tank
Re	Reynolds number
R_m	clean membrane hydraulic resistance
R_p	time-dependent resistance of the growing cake
R_{po}	resistance of the initial foulant deposit
s	cake compressibility
T	absolute temperature Kelvin (K)
t	filtration time
T	operating temperature
T	temperature in K
t	time
t_0	time at which this concentration exits
$t_0\sigma$	standard deviation of the peak
t_f	cake formation time
t_w	time required for the washing
u	fluid circulation velocity
U	fluid velocity
U	overall heat transfer coefficient between the air and the solids
v	velocity
V	bed volume
V	cumulative filtrate volume
v	terminal settling velocity of the solid in a dilute solution
V	total volume of filtrate

v	velocity of liquid through the bed of solid
v	velocity of the slurry flowing
V	volume fraction of each component
V	volume of the tank
V_0	volume required to elute the maximum concentration c_0
$V_0 \sigma$	standard deviation
V_f	volume of filtrate collected during that period
V_m	volume of mobile phase
V_r	volume eluted from the start of sample injection to the peak maximum
V_s	volume of stationary phase
Vs	single stage volume ($=V/N$)
V_w	volume of wash water required
$W_{1/2}$	peak width measured at half peak height
$W_{1/2av}$	average half width of the two peaks
x	concentrations in mol per volume in each of the stream
x	direction along the flow
x	solute concentration in solution (gm/mL)
X	weight fraction of each component
x and x_F	the concentration of the solute in the final and feed solution, respectively
x and x_F	the solute concentrations in the outlet and the feed, respectively
x^*	concentration in the solution which would be in equilibrium with the adsorbent
X_A and X_B	mole fractions of component A and B, respectively
$X_f, X_r, Y_0,$ and Y_1	the weight fractions of solute in the feed, raffinate/solvent and extract, respectively
$X_{f,i}$	mole fraction of component i in the feed liquid
y_A	mole fraction of A
$Y_{p,i}$	mole fraction of component i in the permeate

Greek Symbols

α	area blocked per mass of deposit
α_s	specific resistance of the foulant cake
β	fraction of time that the filter is submerged

Δp	pressure drop across the bed
ΔP_i	change in partial pressure of pure component i across the membrane
ΔP	pressure applied
Δt_{ij}	separation between peaks
ε	voidage
ε	volume fraction of liquid (or void space) in the stage
ε_b	void fraction in the bed
ε_c	void fraction in the cake
η	removal efficiency
θ	angle the disk makes with the vertical axis
μ	fluid viscosity
μ	viscosity of liquid
ρ	density
ρ	liquid density
ρ_o	mass of cake solids per volume of filtrate
$\upsilon \ (=F/A)$	superficial velocity

PREFACE

In 2012 the world chemical and pharma market was € 4068 billion (http://www.statista.com/statistics/263111/revenue-of-the-chemical-industry-worldwide-and-in-the-eu/). The US market revenue of organic chemical industries is about $147 billion (2009) with an annual growth expected to be 11.2%. The contribution of chemical and pharma industries to the overall US economy is about 8.7%. The global biotech market is $289 billion, with annual growth of 10.8% (2009) (http://www.ibisworld.com/industry/global/global-biotechnology.html). The United States is the leading biotech player, with more than 60 billion US dollars of revenue. In 2006, biopharmaceuticals generated approximately 150 billion US dollars of revenue worldwide (2011). Biotechnology products represented 21% of the total $714 billion global market for prescription drugs in 2012, equivalent to $150 billion of sales. In 2010, the sales of industrial chemicals created by using biotechnology in at least one step of the production process equalled €92 billion globally, and this is expected to increase to €228 billion by 2015. The total biotech industry size in 2013 in India is US $4.3 billion (biopharma taking up 64% of the share). It ranks second in Asia and 12th globally (http://www.ibef.org/industry/biotechnology-india.aspx). The above statistics emphasize the fact that chemical/pharmaceutical and biochemical/biotechnology industries play a very important role in the global economy, contributing sizably to the GDP, and are expected to grow in a healthy manner for a long time to come.

While the product is manufactured in a reactor/bioreactor/fermentor, it is recovered and purified in subsequent unit operations, which could be several in numbers. The economy of a manufacturing process is determined by the cost effectiveness of the downstream operations. The downstream processing steps will vary depending upon whether the product is a bulk or a high value chemical. There is some overlap between the downstream processing steps of a chemical and a biochemical process, although there are a few steps that are unique to the biochemical process, such as chromatography.

This book discusses downstream and unit operations practiced in chemical and biochemical industries. The commercial scale of operation of these two industries, as indicated above from the financial, is very large, and so the efficiency and cost involved in downstream processing determines the profitability of these industries.

Chapter 1 introduces the various principles involved in downstream, cost factors, and other issues. Chapter 2 discusses diverse chemical and biochemical industrial processes with special focus on downstream unit operations. Chapter

3 focuses on particle size reduction, bacterial cell breakage, and recovery of intracellular material. The latter is not relevant if the product is released by the microorganisms into the extracellular medium. Chapter 4 deals with the isolation or removal of solids from a solution. Here the solids could be dead biomass or impurities. Chapter 5 discusses various product recovery techniques while chapter 6 deals with product purification/enrichment techniques. Product polishing, stabilisation, and finishing techniques are discussed in chapter 7. A manufacturing plant has several utility services, which are discussed in chapter 8. The future trends and research opportunities in the area of downstream are discussed in chapter 9. A few basic fundamental concepts of chemical and biochemical engineering are discussed in chapter 10. The book includes several problems at the end of each chapter, which will help the reader to assimilate the material. The book contains several line diagrams and mathematical formulae that could be used for design purposes.

—*Mukesh Doble*
October 15, 2015

ABOUT THE AUTHOR

Mukesh Doble, PhD, FRSC
Mukesh Doble, PhD, is a Professor and Head of the Department of Biotechnology at IIT Madras in Chennai, India. He has previously worked for 20 years at Imperial Chemical Industries (ICI) and General Electric (GE) Technology centers. His areas of interest are drug design, biomaterials, bioreactors, and bioremediation. He holds BTech and MTech degrees in chemical engineering from IIT, Madras, India, and a PhD from the University of Aston, Birmingham, UK, and his postdoctoral work was performed at the University of Cambridge, UK, and Texas A&M, Texas, USA. He has authored or coauthored 240 technical papers, seven books, including the books *Cyclic Beta-Glucans from Microorganisms: Production, Properties and Applications; Drug Design, Basics and Applications; Green Chemistry and Processes; Biochemical Engineering; Biotreatment of Industrial Effluents; Biotransformations and Bioprocesses*; and *Homogeneous Catalysis: Mechnanisms and Industrial Applications*. He has filed six patents. He is a fellow of the Royal Society of Chemistry, London, and a recipient of the Herdillia Award for Excellence in Basic Research from the Indian Institute of Chemical Engineers and was named the Dow Professor M. M. Sharma Distinguished Visiting Professorship in Chemical Engineering at Institute of Chemical Technology, Mumbai, India. He is on the editorial board of the journal *Chemical Engineering* and a member of the American and Indian Institutes of Chemical Engineers.

CHAPTER 1

DOWNSTREAM PROCESSING PRINCIPLES, ECONOMICS AND ISSUES

CONTENTS

1.1 INTRODUCTION

A chemical (Figure 1.1a) or a biochemical (Figure 1.1b) process consists of an upstream, reactor and downstream sections. The upstream section in a chemical plant will consist of the raw materials preparation. Whereas, the upstream section in a biochemical process consists of facilities for the preparation of micro-organism and media, sterilization of raw materials

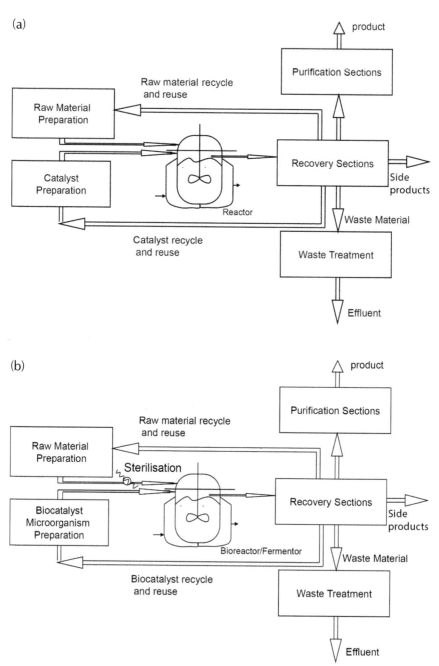

FIGURE 1.1   (a) Chemical process flow sheet; (b) Biochemical process flow sheet.

and inoculation of the micro-organism. The reactor section is the fermentor or a reactor/bioreactor and the associated machinery. An aerobic fermentor may contain air purification, compression and injection systems. The product is prepared in the reactor section. The downstream in a biochemical or chemical process section may consists of the following:

1. separation, recovery, and recycle of micro-organisms, enzymes, catalysts (metal, heterogeneous or homogeneous catalyst, if present) or raw materials,
2. recovery of the product,
3. purification of the product to the required degree, and
4. effluent treatment and disposal.

Downstream is where the product is recovered, purified, and stabilized. Recovery of side product is also carried out in the downstream section.

In addition utilities such as, air, steam, coolant, and hot oil are the other sections, which are also associated with a manufacturing facility. Figures 1.1a and 1.1b show a typical process flow sheet. Sections 1.2, 1.3, and 1.4 may be called the downstream processing section and effluent treatment is generally considered as a separate unit.

Downstream plays a crucial role in arriving at a pure product in an economical way. Purity becomes very crucial for pharmaceutical and health care products. A complex downstream section adds to the final selling price of the product and hence, its competitive nature in the market. Currently, bulk chemicals, pharmaceuticals, antibiotics and food products are manufactured using biochemical route. The process could be based on fermentation or a biocatalyst such as, a pure enzyme or a whole cell. Chemical route is still competitive than biochemical route for bulk chemicals, but the advantages offered by the latter over the former is tremendous in several areas (such as, milder conditions, better product quality, green approach and less effluents, etc.).

Chapter 2 introduces several products manufactured through chemical and biochemical routes, with main focus on the downstream. A typical downstream consists of removal of insolubles, isolation of product, its purification and polishing. The initial removal of insolubles from the mother liquor includes filtration and centrifugation. Isolation of products involves removing the desired product from a very dilute product liquid (may be from 5 to 15%) and it includes adsorption, solvent extraction, distillation, etc. This step is meant to concentrate the product. The third step involves the purification of the product and the operations include chromatography, electrophoresis, and crystallization (or distillation if the product is stable at high temperature). Depending upon the product purity desired these operations are repeated many times. The final step will include crystallization, lyophilization, stabilization of the product and drying. Steps such as, crystallization, lyophilization, and drying are required if the desired product is in the solid form.

After fermentation, if the product is extracellular in nature (i.e., if the desired metabolite diffuses out and accumulates in the fermentation broth), then the biomass is collected after the reaction and is disposed and the liquid will be processed for the product. Whereas, if the desired product is intracellular in nature (i.e., if the product accumulates inside the cell), then the biomass has to be collected after the reaction, the cells have to be broken and the intracellular material needs to be extracted from this liquor. So the intracellular products will include few more downstream steps such as, cell harvesting, cell breakage and removal of cell debris and other unwanted proteins. This issue is not relevant in a chemical process.

1.2 VARIOUS DOWNSTREAM OPERATIONS

Figures 1.2a and 1.2b show the possible set of unit operations involved in the downstream for recovery of extra and intracellular product respectively from a biochemical reaction. If the product is an intracellular material then the cells have to be harvested and they have to be disrupted to liberate the product. This disruption includes mechanical, chemical and biochemical methods. Removal of cell debris is also a difficult step. Several primary, secondary and final recovery and purification steps can be adopted depending upon the concentration of the product, purity desired, its physico-chemical properties, and properties of the other impurities in the medium and cost of the recovery operation.

The various unit operations in the downstream can be listed based on the physical–chemical principles as shown in Table 1.1.

A typical design of downstream unit includes determining (i) the cycle time for the operation, (ii) the deciding on the operating parameters such as,

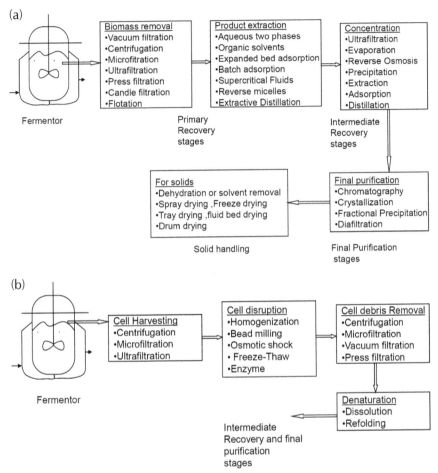

FIGURE 1.2 (a) Various downstream processing alternatives for extracellular products from fermentation; (b) Various alternatives for recovering intracellular products from fermentation.

temperature, pH, flow rate, pressure, agitation intensity and (iii) the finally the hardware details such as, size, diameter, height of the unit and material of construction, etc. There are several issues involved in the design of the downstream section, which includes availability of design equations (detailed or superficial design), parameters/data to perform the calculation, conversion factors, design safety margin, idealization of the design equation and vendors who will be supplying the equipment. Discussions with vendors may throw light on the hardware details.

TABLE 1.1   Various unit operations in the downstream listed based on the physical–chemical principles

Unit operations	Physical–chemical principle
Filtration (screens or membrane)	Particle size
Size reduction	Particle size
Sedimentation	Particle size
Gel permeation chromatography	Molecular weight/size
Centrifugation/cyclone separation	Density
Denaturation/precipitation/crystallization	Temperature
Dialysis	Membrane–ion-interaction
Electrodialysis	Membrane–molecule interaction + Electric forces
Affinity chromatography	Affinity between ligand and protein/enzyme
Reversed-phase chromatography	Hydrophobic interactions
Ion exchange chromatography	Ionic forces
Reverse osmotic membrane	Osmotic pressure
Pervaporation	Membrane–molecule interaction
Liquid–liquid extraction	Partition coefficient
Adsorption	Surface
Absorption	Bulk (gas into liquid)
Distillation	Vapor pressure/boiling point différences
Drying	Evaporation of vapor
Lyophilization	Sublimation

1.3   COST FACTORS

The most important point to consider is the economics of the downstream operation. Hence knowledge of the equipment (i.e., capital cost) and operating costs are necessary. The total cost for performing certain operation is the sum of fixed costs (capital cost) and variable costs. Fixed costs are the sum of many components and they do not change when the volume of production is changed. Fixed costs can include facilities costs, certain general and administrative costs, and bank interest and depreciation expenses. Variable cost is the expenses associated with producing the chemical. It changes with the volume of production. It includes direct raw materials used, labor cost, transportation, sales commission, and advertisement expenses. The variable unit cost is the cost associated with producing one unit (which is equal to total variable cost divided by number of units produced). The cost of an unit or product includes production cost, which is made up of direct manpower (indirect manpower may have to be included in the overhead or fixed cost), raw materials, electricity, transport, rent, water, machinery, equipment, tools and others.

If the cost of an equipment is known then the approximate cost of a larger or smaller sized equipment can be as shown below (This is known as 0.6 rule).

$$\text{Cost } 2 = \text{Cost } 1 \, (\text{Size } 2/\text{Size } 1)^a$$

where a ~ 0.6 (can vary between 0.5 and 1.0).

The cost of the equipment not only includes the direct cost of the vessel but also other factors such as, installation, piping, electrical fittings, insulation, contractor fee's, etc., as shown in the Table 1.2. So the direct

TABLE 1.2 Multiplication Factors for Various Installations

Total plant direct cost (TPDC)	Multiplication factors
Equipment purchase cost	PC
Installation	0.5×PC
Process piping	0.4×PC
Instrumentation	0.35×PC
Insulation	0.03×PC
Electrical	0.15×PC
Buildings	0.45×PC
Ground/yard improvement	0.15×PC
Auxiliary facilities	0.5×PC
Total plant indirect cost (TPIC)	
Engineering	0.25×TPDC
construction	0.35×TPDC
Total plant cost (TPC)	TPDC+TPIC
Contractor's fees	0.05×TPC
Contingency	0.10×TPC
Direct fixed capital	TPC + Contractor's fees + Contingency

fixed capital may be 2 to 3 times the actual cost of the equipment. Cost of a micro/ultra filtration membrane is ~30,000 US$ for a 20 m² membrane area and it increases to 110,000 US$ for a membrane unit with 100 m² area (1998). Cost of a 1000 L column will be 100,000 US$ (1998). The vendors will provide the actual data depending upon the requirement, operating conditions and material of construction.

There are several terminologies related to profitability analysis which one would need to know and they are gross margin, return on investment (ROI), payback period, gross profit, net profit, depreciation, amortization and depletion. The definitions of these are as follows:

Gross Margin = Gross profit/Revenues
Return on investment = Net profit per year × 100/Total investment
Pay back period = Total investment/Net profit per year
Gross profit = Annual revenues – Annual operating cost
Net profit = Gross profit – (Income tax + Depreciation)

Amortization is the term used to allocate cost of intangible assets, such as, patents, copyrights, trademarks, and franchises.

Depletion is the term used to record the cost of natural resources extracted from the earth.

In addition, there are other concepts in profitability one needs to know and they are described below.

1.3.1   DEPRECIATION

Depreciation is the expense incurred which is spread over a period of time and it spans over the life time of the equipment. The expenses could be the cost of major equipment. Depreciation is referred to as cost recovery, mainly the cost of plant assets. Depreciable cost includes buildings, equipment, vehicles, computers, furniture and fixtures. This cost includes the purchase price, sales tax, shipping and installation costs, and direct incidental costs.

The most commonly used method for calculating depreciation is the straight line method. This method assumes that an asset should be depreciated in equal amounts each year until it reaches its scrap value (i.e., residual value when the asset has no more use). The number of years an asset lasts depends on the item, for example, some equipment may last for 5 years and some for 10 years. The formula for calculating depreciation is based on the following method:

Rate of Depreciation = (Original Value – Residual Value)/Expected Life

For example, a filter is purchased for $5,00,000 and you expect to use it for five years and if the residual value is estimated at $ 50,000. Then,

Rate of Depreciation = [5,00,000 – 50,000]/5 = $90,000

1.3.2 NET PRESENT VALUE

Net present value (NPV) compares the value of a dollar (or rupees) today to the value of that same dollar (or rupees) in the future, taking into account the inflation and returns. If the NPV of a project is positive, it should be accepted. If NPV is negative, the project should be rejected because cash flows will also be negative. For example, if you want to purchase a chromatograph you would first estimate the future cash flows that this equipment would generate, and then discount those cash flows to the present cost and estimate whether you will end up making a profit or loss. The future cash flow can be calculated from the discounted cash flow.

1.3.3 DISCOUNTED CASH FLOW

Discounted cash flow (DCF) analysis is a method used to calculate the present cash flow projections and discount them (most often using the weighted average cost of capital) to arrive at a present value, which is used to evaluate the potential for investment. If the value arrived at through DCF analysis is higher than the current cost of the investment, the opportunity may be good.

For example, if the cost of a chromatograph is CC and profit incurred due to it is CF_1 in year 1, CF_2 in year 2, ... and CF_n in year n then,

$$DCF \text{(profit discounted to the present value)} = \frac{CF_1}{(1+r)^1} + \frac{CF_2}{(1+r)^2}$$

$$+ \frac{CF_3}{(1+r)^3} + ... \frac{CF_n}{(1+r)^n}$$

CF = cash flow, and r = discount rate. If CC – Profit DCF > 0, then we will be incurring a loss due to that investment and so it is not prudent to buy the equipment. If CC – Profit DCF < 0, then we will be incurring a profit due to the purchase of the equipment. Here, we assume that after n years there is no resale value for the chromatograph. If we have a resale value after n years, we may include it as a cash flow from year n. Generally, for Indian scenario we may consider r = 0.10 (i.e., 10%). In addition, we may be incurring some expenditure every year to repair or maintain the unit. So we need to perform a similar CF out and include it in the DCF calculations.

There are many variations when it comes to what you can use for your cash flows and discount rate (r). The purpose of DCF analysis is to esti-mate the money one would receive from an investment and to adjust for the time value of the money. DCF models have their advantages and disad-vantages. If inflation is 6%, then the value of the money would halve every 12 years. For example if one expects an equipment to give an income of $30,000 in the 12th year, then it is equal to $15,000 today. This technique can also be extended to R&D expenses, profits and gains, etc.

1.3.4 PAY BACK PERIOD

The length of time required to recover the cost of an investment (land cost, equipment, facilities, licensing, etc.) is known as pay back period. Generally, companies prefer shorter pay back period than longer periods. If all other things are assumed equal, then the better investment is the one with the shorter payback period. For example, if cost of setting up a plant is US$100,000 and it is expected to return $20,000 annually, the payback period for the investment will be $100,000/$20,000 = 5 years. There are two main problems with the payback period method and they are: (i) it ignores any benefits that may occur after the payback period and hence, it is not a measure of profitability; and (ii) it ignores the time value of money (i.e., change in the value of money with time). Because of these reasons, net present value, internal rate of return or discounted cash flow are generally preferred for capital budgeting.

1.3.5 INTERNAL RATE OF RETURN

Higher a project's internal rate of return (IRR), the more desirable it is to undertake the project. IRR can be used to rank several prospective projects. The project with the highest IRR would be considered the best and selected first. IRR is sometimes referred to as "economic rate of return (ERR)."

During the design of a downstream operation one should have a reasonable knowledge about the cost of the product, what is the acceptable product quality, where is the product in each of the streams, where are the impurities coming into the product, physico-chemical properties of the product and other major impurities, other alternative approaches for the operation and the comparison of the cost of these alternative approaches.

1.4 MASS AND ENERGY BALANCE

One needs to perform the mass balance across each of the unit operation (Figure 1.3) to evaluate the efficiency of the process and losses involved. Loss of product means decrease in yield/recovery or loss of profit. At steady state, the amount of material(s) entering through various streams should be equal to the amount of material(s) leaving the unit through various streams as shown below. This equation is called the mass balance.

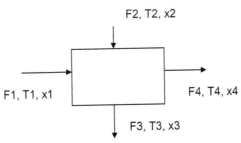

FIGURE 1.3 Mass inputs to and outputs from an unit.

$$F1\ x1 + F2\ x2 = F3\ x3 + F4\ x4 \tag{1.1}$$

where Fs are the volumetric flow rates and x are the concentrations in mol per volume in each of the stream. F1 and F2 are the streams entering the unit and F3 and F4 are the streams leaving the unit.

If there is a reaction taking place inside a unit leading to the conversion of one component to another, then there will be a decrease in the first component and a corresponding increase in the second component as per the stoichiometry of the reaction. The mass balance for component one will be:

$$F1x1 + F2\ x2 = F3\ x3 + F4\ x4 - Vr \tag{1.2}$$

where V is the volume and r is the rate of reaction (leading to disappearance of component 1).

If there is an accumulation of material inside the unit, especially during the start up, then one more term needs to be included in the model which will be a function of time as

$$F1\ x1 + F2\ x2 = F3\ x3 + F4\ x4 + V\ dx/dt \tag{1.3}$$

where x is the concentration inside the unit, which varies as a function of time. The last term corresponds to accumulation. Once the operation reaches steady state, then dx/dt will reach zero. Similarly, when feed(s) to the unit is stopped then concentration inside the vessel may decrease (then dx/dt will be negative). If there is a chemical reaction inside the unit then concentration inside the unit may decrease and a new product may be formed.

Similar to mass balance, heat balance of the unit can also be developed (Figure 1.4) as,

$$\text{Heat input} + \text{Heat generated} = \text{Heat output} + \text{Heat loss} \tag{1.4}$$

The term heat generated could be due to exothermic reactions inside the vessel/reactor (correspondingly there could be heat decrease inside the unit if there is an endothermic reaction). Heat is also generated due to addition of chemicals or mixing of the streams entering the unit, as well as the heat input by the heating fluid or steam. Heat loss includes heat lost to the ambient or heat removed by the cooling system.

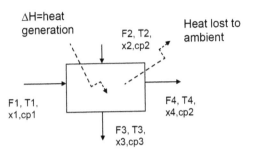

FIGURE 1.4 Heat inputs to and outputs from an unit.

$$\text{Heat input by a fluid entering the unit} = \begin{array}{l} \text{Mass flow} \times \text{Specific heat} \times \\ \text{Temperature difference between} \\ \text{data and baseline temperature} \end{array}$$

(1.5)

Multiple stage operations may involve several reactors and several separators in series as shown in the Figure 1.5. One could perform a mass balance for the entire train. The yield in one unit may be high, but when yield from many units are multiplied the overall yield of the entire set of unit operations may drop down to much smaller value as seen in Figure 1.5. For example, although the fractional yield (Y) in each of the reactor is 0.95 and the efficiency of each separating unit (η) is 0.98, the overall fractional yield of the entire 3 reactors cum 3 separating units drop down to 0.8 ($= 0.95^3 \times 0.98^3$), leading to 0.2 waste ($= 1 - 0.8$). This also means that the total load on the effluent treatment system will be 0.2. Also each of the reactors is producing 0.95 of the desired product, which means the

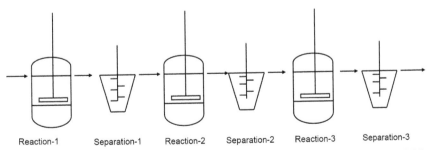

Reaction-1   Separation-1   Reaction-2   Separation-2   Reaction-3   Separation-3

FIGURE 1.5 The concept of yield and purity in a recovery train (Y in each reactor =0.95, h in each separator = 0.98).

remaining material is side or wasteful product. So total waste produced by these three reactors will be it 0.14 (= 1 − 0.95³). Loss of products in the combined separating units is 0.16 (= 1 − 0.98³). This example indicates that the yields and separation efficiencies have to be very high in each unit.

1.5 SAFETY

Safety is an important component while designing downstream units. Instead of adding safety features in a unit operation, it is much more prudent to design unit operations which are inherently safe. Principle considerations in inherent safety are to avoid processing or using toxic, flammable or environmentally hazardous materials; reducing the inventory of hazardous material; reducing the potential for surprise (which means use of known chemicals and trusted processes) and separating people from chemicals and solvents. There are several key words that are used to develop inherently safe process and operations and they are listed below (Kletz, 2006).

- Intensification: Use concentrated chemicals so that the unit sizes will be small.
- Substitution: Substitute toxic substances with less toxic ones.
- Use non-hazardous reactions and processes.
- Attenuation-reduction of the impact. The design should be such that in case there is an accident the impact felt should be minimal. This means the buffer should take care of the accident.
- Limitation of effect: Changing designs or process conditions rather than by adding protective equipment which may fail. For example, it is better to prevent overheating by using a fluid at a lower temperature (steam) rather than using a hot oil and relying on a control system to prevent over heating. The steam temperature is limited (based on the steam pressure) while the hot oil temperature is not limited, hence, one may require safety checks to control its temperature.
- Contain/enclose/reinforce-enclose or contain any dangerous operation. Separate human operators from the vessels.
- Error tolerance: The equipment should be able to tolerate errors in design and errors in process operation. It should not be too sensitive, that small changes may lead to run away conditions.

- Avoid knock-on effects: Problem in one unit should not affect other units downstream.
- Ease of control: The unit should be easy to control and operate by human.

Use non-toxic chemicals and solvents – use safer chemical in place of a more hazardous one. It may be possible to replace flammable and low boiling solvent with non-flammable and high boiling one or toxic chemical with non-toxic one. It is necessary to evaluate not only the property of the substance but also the volume that is handled. Changing to a safer reagent may be a better solution than using a hazardous chemical and incorporating several safety checks and control alarms and trips to prevent the hazards. An example of this principle is the use of air, instead of hydrogen peroxide in oxidation reactions or the use of hydrogen instead of hydrazine for reduction reaction. Another example is use of magnesium hydroxide slurry instead of concentrated sodium hydroxide solution to control pH. The former is less corrosive than the latter. Inherent safety could be achieved by using milder alkalis or acids in reactions instead of stronger alkalis and acids.

Many processes have large inventories of toxic or flammable materials, because the percentage conversion is low or the rate of reaction is slow. This problem also arises in solvent extraction or gas liquid absorption. These are a major source of hazards and effort should be made in reducing the inventory. This thought should be given at the research stage and it may not be possible to rectify the problem during process development or pilot plant stage. It is worth remembering that the best way of avoiding a leak of hazardous material is to use so little that it does not matter even if it leaks out. Extraction of the active ingredient from the fermentation broth requires large volumes of chemicals. If the partition coefficient of the solute in the extraction solvent is high, then one requires low quantities of the solvent. Identifying the best solvent can be easily carried out in the research and development stage.

The temperature and pressure at which the process operation is carried out will affect the inventory, for example, increasing the pressure in a phosgene plant reduces the inventory to one tenth of the original level. The inventory of the reaction is affected by the concentration. One needs to think whether the concentration can be increased safely, bearing in mind potential temperature rise due to the heat of reaction as the reaction is carried out at concentrated condition. In biochemical process, increasing concentration may decrease the rate of reaction, if the biocatalyst or organism is inhibited by substrate. The number of stages in the reaction will have an impact on the overall inventory especially if intermediates are isolated at every stage. Isolation at each stage increases hazard.

The two challenges in the concept of inherent safety are: (i) Measuring the degree of inherent safety so that it allows comparisons of alternative designs. The alternate designs may or may not increase safety but may simply redistribute the risk. (ii) The second challenge is that inherent safety needs to be implemented early during the process development stage. Later it becomes difficult or impossible. The technical challenges of inherent safety may require R&D effort.

1.6   GREEN CHEMISTRY APPROACHES

Telescoping several unit operations into one single step is one of the best principles advocated by green chemistry (Doble and Kumar, 2007). This approach could help in reducing the wastage due to handling in each stage and also use of many solvents.

Solvents are used in liquid-liquid extraction and chromatography. The choice of solvent, for a process, involves a number of considerations, including toxicity and compatibility with the other raw materials. Another aspect that needs to be kept in mind is its boiling point. Imagine in a particular reaction, acetone is used as the solvent. The heat of reaction of the process is such that an uncontrolled addition of one of the reactants, or loss of cooling, could lead to vigorous boiling of the batch, over-pressurization of the reactor and flooding of the condenser with acetone. If acetone is changed to toluene then the reaction mixture will be one with a boiling point sufficiently high to overcome all the above-mentioned possible hazards. The ambient average temperature, and the highest ambient temperature reached during summer months are also important parameters during the selection of the solvent, more so in equatorial regions, since higher ambient temperature leads to higher vapor pressure and hence higher vapor concentration in the atmosphere.

Using fewer chemicals means less waste and also as far as possible the solvents should be recycled. If too many solvent are used then one may end up with solvent mixtures instead of pure solvents, and the solvent mixtures may require complicated distillation assemblies for separation and purification. If the solvent is miscible in water, then the aqueous stream will contain solvent and there will be difficulty in the effluent plant. So using minimum number of solvents and water immiscible solvents are better.

1.7 SCALE-UP

The separation and purification studies are initially carried out in the lab scale and then they are translated up to pilot plant level. A large number of processes fail to reach commercial scale because of the problems they face during scale-up (Figure 1.6). The science of scale-up for biochemical processes encompasses not only knowledge of biochemistry and biotechnology but several branches of chemistry and engineering including organic and physical chemistry, process chemistry, chemical engineering, and fluid mechanics. Issues relating to thermodynamics and hydrodynamics also become important during scale-up, and they have an impact on the system performances. The operations can be performed in batch or continuous mode. Maintaining sterility in the large vessel, achieving the same degree of mixing of various liquid and solid components in the large scale and overcoming the heat and mass transfer

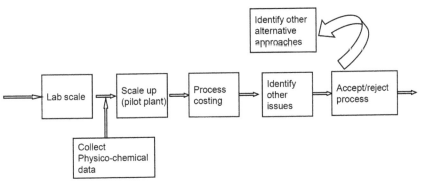

FIGURE 1.6   Various steps during scale-up of a downstream process.

problems are some of the challenges during scale-up. Scale-up can also lead to extended batch cycle time. For example, a filtration, which may take minutes in the lab, may take hours in the pilot scale due to the generation of impurities, which may change the characteristics of the filtrate and prevent free flowing of the liquid. Poor precipitate formation may hinder the settling of solids.

In addition, scale-up of biotechnological processes pose extra problems. Shear fields experienced by the micro-organism in a fermentor are not the same at all scales. Stability of the biocell, mold and enzyme due to mechanical agitation and collision with air and/or gas bubbles is another issue that is unique in biology and should be addressed.

During scale-up of vessels the surface-area-to-volume ratio decreases and hence similar rate of heat input and removal cannot be achieved at all scales. A volumetric scale-up of 10,000 means a reduction in surface-area-to-volume ratio by 21.5. So to input or dissipate the same amount of heat as in small scale, the heat transfer rate must be increased in proportion to the vessel size. This can be achieved by including additional heat transfer area such as, internal coils or higher heating or increasing the medium temperature. The latter can lead to increased temperature of the vessel surface, with the possibility of localized overheating or formation of tarry or polymeric material. Temperature gradient inside the vessel may alter the growth characteristics of the organism.

The minimum rpm required for suspending solids is a function of particle diameter,

$$N_{rpm} \, \alpha \, dp^{0.2}/D^{-0.85} \tag{1.6}$$

and the power required by the agitator per unit volume of the liquid is proportional to $D^{-0.55}$. dp is the particle or solid diameter and D is the agitator diameter. So as the vessel diameter increases power input to the liquid decreases.

If the process contains two immiscible fluids then one may be dispersed in the other. Then, the average emulsion droplet size, d is proportional to $N^{-2.56} \, D^{-4.17} \, T^{1.88}$ and to achieve uniform dispersion, agitator rpm N should be proportional to $D^{-2.15}$. T is the tank diameter.

Typical mixing times in small scales could be about 10 s, while the same in the large plant scale could be of the order of 100 s. Blending time, t α TH/ND^2, and in geometrically similar tanks mixing time is proportional to 1/N.

In scaling-up of many homogeneous stirred systems, the rate of heat transfer controls the design of heat transfer equipment, that is, h α u^m D^{m-1}, where h is the heat transfer coefficient and u is the fluid circulation velocity. m is generally of the order of 0.6714. A scale-up based on equal heat transfer coefficient is desirable when pilot plant studies indicate that the heat transfer resistance increases slowly over a period of time, due to the formation of polymeric or tarry deposits along the heat transfer area, caused by the temperature-sensitive nature of the material.

For highly temperature sensitive material, it is necessary to have same rate of heat transfer in the large scale unit to avoid material degradation, that is, h2/h1 = D2/D1, and for geometrically similar vessels,

$$N2/N1 = (D2/D1)^{(2-2\ m)/m}$$

where h is the heat transfer coefficient and N is the rpm.

There are many issues that need to be considered during scale-up and they are not the focus of this book.

1.8   ENVIRONMENTAL ISSUES

Finally the waste (solid, liquid, gaseous) that is generated in the plant needs to be processed and disposed, so that the properties of the waste (BOD, COD, suspended solids, dissolved solids, pH, DO, metals, etc.) are within the stipulated limits of the local pollution control board. The following points with respect to environmental impact that need to be considered by the process designers while designing their process are:

- Use chemicals that are already available and that are being used in the current plant, so that new effluents will not be generated. New effluents require new treatment strategies.
- Try some new separation chemistry.
- Use of One-pot reactions can decrease waste. Performing the process in too many vessels leads to large quantity of effluent arising due to vessel cleaning and washing. If multiple steps are performed in the same vessel the amount of effluent generated will also be less.

- Use Smaller Reaction Vessels. This can lead to less quantity of effluent.
- Recover, Recycle, and Reuse. All process solvents, catalyst and secondary reagents if they can be recovered, recycled and reused, then the effluent generated will be less.
- Convert "Waste" into "Product." If waste cannot be avoided then a process could be developed to convert the waste into useful product, which can be marketed.
- Minimize solid and liquid effluents by optimizing the process yield. An optimum process will require minimum amount of chemicals and reagents.
- Don't use too many solvents, which may lead to solvent mixtures. Chromatography may require gradient separation, which may involve more than one solvent leading to solvent mixtures in the downstream. Separating mixtures of solvent in the effluent is not easy.
- Using one chemical to break binary azeotrope leads to combinations of solvents in the effluents, which cannot be removed.
- Avoid washing of vessel, since that generates wastes. Use final rinse solvent as primary wash solvent.
- Some products can withstand contamination; if so, do not over purify it.
- Use solvents of low water solubility; else they will get carried with the aqueous stream into the effluent.

1.9   UTILITIES

Any process plant will require steam, hot oil and hot water for heating the reactants, and also will require chilled water and coolant for cooling the products or reactor contents. In addition, a process plant would also require air, nitrogen, oxygen and possibly other gases too. All these are called utilities and a typical chemical or biochemical process plant will have facilities to produce them. The cost of manufacturing such utilities also need to be included in the selling price of the final product. Many different utilities would also mean several facilities for their manufacture. Some gases may be purchased in bulk. Details of utilities are given in Chapter 8.

KEYWORDS

- **depreciation**
- **net present value**
- **net profit**
- **pay back period**
- **safety**
- **scale-up**

REFERENCES

1. Bioseparation and Bioprocessing (2nd Ed.) 2nd Volume Set, ed. Subramanian Ganapathy, Wiley-VCH, (09–2007).
2. Bioseparations – Principles and Techniques, B. Sivasankar, Prentice Hall of India, N Delhi, 2005, pp. 280.
3. Bioseparations: Downstream Processing for Biotechnology (Hardcover), by Paul A. Belter (Author), E. L. Cussler (Author), Wei-Shou Hu (Author), Wiley-Interscience; 1st edition (February 8, 1988).
4. Downstream Processing of Natural Products. A Practical Handbook. Edited by M. S. Verrall, Wiley (August 15, 1996), pp. 372.
5. Downstream Processing of Proteins: Methods and Protocols (Methods in Biotechnology) (Methods in Biotechnology) (Hardcover) by Mohamed A. Desai (Editor), Humana Press; 1 edition (March 31, 2000), pp. 240.
6. Handbook of Downstream Processing. Edited by E. Goldberg, Blackie Academic & Professional, London., 1997.
7. http://www.sciencedirect.com/science/book/9780123725325.
8. Kletz, Trevor (2006). Hazop and Hazan (4th Edition ed.). Taylor & Francis. ISBN 0852955065.
9. Mukesh Doble; Anil Kumar, Green Chemistry and Engineering, Elsevier, NY (Jun 2007).
10. Protein purification: Design and scale-up of downstream processing, Scott M Wheelwright, Wiley Blackwell; New Ed edition (23 May 1994), pp. 244.

PROBLEMS

1. Antibiotic production consists of fermentation followed by filtration of the biomass. The antibiotic, which is in the filtrate, is extracted using a solvent; which is stripped in a distillation column. The original broth contains 15 wt.% of biomass, 25 wt.%

of penicillin and rest mother liquor. The solids retain 5 wt.% of the solution in the filter Two parts of solvent by weight is added for every part of the antibiotic in the mixer extractor. Solvent extraction process is only 97 wt.% efficient (i.e., 3 wt.% of the antibiotic is left behind in the mother liquor). The solvent carries 1 wt.% of the penicillin away during stripping operation. What is the efficiency of this downstream process with respect to penicillin recovery?

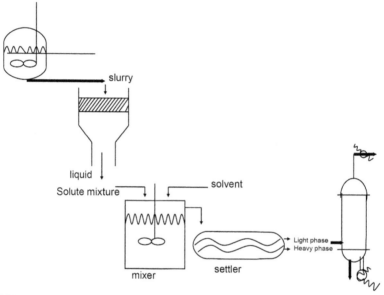

PROBLEM 1.1

2. A fermentation broth contains 15 wt.% of dead mass and rest liquid, the latter needs to be recovered fully. The annual production of the broth is 20,000 kg. The liquid can be sold at a profit of Rs. 100/kg. If a simple filter is used the solids retain 10 wt.% of the liquid, while if a centrifuge is used the solids retain only 2 wt.% of the liquid. Cost of a filter is Rs. 2,00,000 while the cost of a centrifuge is Rs. 2,50,000. The annual maintenance plus operating costs of the filter is Rs. 50,000 and that of the centrifuge is Rs. 80,000 per annum. If you assume that the life of both the equipment is 4 years and there is no resale value for them, suggest the correct filtration equipment with reason. Assume all other factors are the same. The discount factor for the money is 10%. Assume that the equipment will be purchased at the beginning of year 0, and the profit on sales will be received in the beginning of year 1, 2, 3 and 4. Also assume that all the operating costs will be accrued in the beginning of year 1, 2, 3 and 4.

3. A chromatographic separation can recover 90% of the desired protein and it requires 1000 liters of methanol. Addition one or more chromatograph in the downstream can increase the recovery to 98% and it will require 500 liters of ethyl acetate. Draw the flow sheet of the various downstream units including solvent recoveries. Discuss this flow sheet with respect to the principles of Green chemistry.

4. Calculate the heat required to heat a fluid of 1600 liters from 30 to 80°C and vaporize it at its boiling point (80°C). Assume its specific gravity as 1 gm/mL, specific heat as 85 kcals/gm. Estimate the quantity of heat transfer liquid required to perform this operation, which is at 100°C.

5. Air from a fermentor contains 5% of CO_2, which we want to remove. Suggest various removal strategies and identify which one could be economical.

CHAPTER 2

IMPORTANCE OF DOWNSTREAM IN INDUSTRIAL PROCESSES – EXAMPLES

CONTENTS

This chapter deals with various industrial processes related to the manufacture of chemical and biotechnology products with a special focus on the downstream. An overview of the existing downstream facilities will help the engineers while designing new manufacturing facilities. These examples are selected so that they cover most of the downstream operations. Detailed discussions of the downstream operations are provided in the subsequent chapters.

2.1   BEER PRODUCTION

Beer is produced by fermentation using *S. cervevisae*. The plant includes not only the fermentor for preparing the beer but also several downstream units such as, filters (screen), distillation columns and dryers (Figure 2.1). The top fraction of the final distillation column gives the high purity alcohol. Some of the major Companies involved worldwide in production of Ethanol (2012) are Archer Daniels Midland (Cedar Rapids) with annual capacity of 1070 million gallons per year (http://www.admworld.com), VeraSun Energy (aurora) with annual capacity of 120 million gallons per year (http://www.verasun.com), New Energy Corporation (South Bend) with annual capacity of 102 in million gallons per year and Hawkeye Renewables, (Fairbank) with annual production of 100 million gallons per year (http://www.hawkrenew.com)

Raw materials constitutes upto 70% of the cost of ethanol production. Cane sugar is extensively used in Brazil and India, while corn is used in United States. Brewing produces dilute concentrations of ethanol in

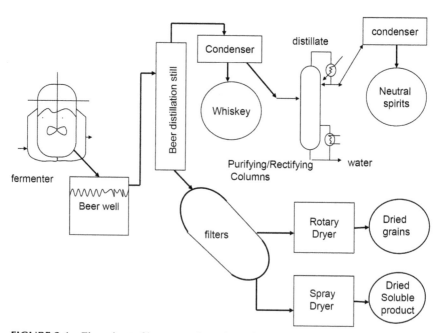

FIGURE 2.1   Flow sheet of beer manufacturing unit.

water. Concentrated ethanol is toxic to yeast and the most ethanol-tolerant strains can survive up to 15% ethanol (by volume), and optimal temperature from 50 to 60°C. A batch fermentation is anaerobic and the cycle lasts nearly 30–40 h.

Separation of ethanol from fermentation broth is performed by distillation, which is energy intensive and contributes to more than 50% of the total energy consumption of the plant. There are two types of distillation processes, the continuous-feed and batch distillation systems. Ethanol and water form an azeotrope (a constant boiling mixture), with a boiling point of 78.1°C. With the help of an organic solvent, such as, pentane or gasoline (entrainer), 100% of alcohol can be produced, which is added to break the azeotrope. This mixture is fed into a distillation column, which divides it into top and bottom products, the latter is pure alcohol. The distillate of this column is fed to a third column, which distils out the solvent, leaving the bottom product as a mixture of alcohol and water. This bottom product is returned to the first alcohol-water column.

Adsorption is adopted to remove trace amount of water from the alcohol. A column is packed with finely ground, dry cornmeal, which is a selective absorbent of water from ethanol/water vapor. The process uses two parallel columns, with one column used for adsorption, while the other is being regenerated. Regeneration is achieved by passing a hot inert gas through the organic bed to evaporate the absorbed water.

2.2   AMINO ACID PRODUCTION

Amino acids are a multi-billion dollar business (2012), with an annual growth of 2.5% and they are used as animal feed additives (*lysine, methionine, threonine*), flavor enhancers (*monosodium glutamic, serine, aspartic acid*) and in the medical field. Glutamic acid, lysine and methionine (1.25, 0.5 and 0.5 MTa, respectively, 2000) account for the majority, and the first two are made by fermentation; and the last is made by chemical synthesis. L-Aspartic acid and L-alanine are made through enzymatic synthesis. The major producers of amino acids are Japan, US, South Korea, China and Europe. Monosodium glutamate was first marketed as a flavoring material in 1909 and its production has reached almost 1 billion dollar (2003).

Four amino acids namely, L-aspartic acid, L-phenylalanine, L-leucine HCl, and L-leucine are produced in batch mode. These amino acids are primarily used as dietary supplements. L-Aspartic acid and L-phenylalanine are the two main ingredients in the artificial sweetener aspartame. The break-even prices for the L-leucine and L-lysine processes are $13.30/kg and $2.70/kg, respectively. Raw materials represent the largest cost followed by waste treatment and utility cost. 67% of the cost is due to the nutrient media, 12% is due to the NH_4OH used in the ion exchange column and 13% is the cost of the precoat. Figure 2.2 shows the cost break down for the equipment. Reactors contribute to 33% of the cost. Downstream such as, crystallizers, storage tank and filters contribute to 28, 19 and 12% of the total hardware cost, respectively. In the case of L-aspartic acid, and L-phenylalanine, 95% of the raw material cost is due to the media. Once again after the reactor, the cost of crystallizers, storage vessels and filters take up most of the hardware cost (Figure 2.3). Industries producing amino acids are Ham, France by Extraction from proteins hydrolyzates, Konstanz, Germany through Enzymatic resolution and Nanning, China through biocatalysis and fermentation.

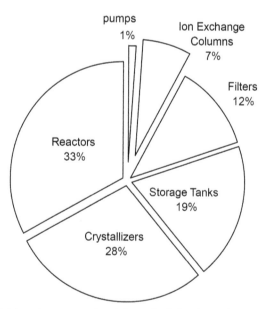

FIGURE 2.2   Total equipment cost breakdown for L-leusine and L-lysine manufacturing facility.

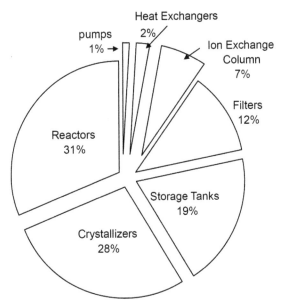

FIGURE 2.3   Total equipment cost breakdown for L-aspartic acid L-phenylalanine.

The amino acid (after its production) is recovered from the broth and purified using several downstream steps which includes reverse osmosis, and chromatography. Reverse osmosis is a filtration process that uses pressure to force a solvent through a filter that retains the solute on one side and allows the pure solvent to pass to the other side.

Chromatography is the main step to achieve highly pure single amino acids. The technique includes ion exchange and high pressure liquid chromatography. In the former the solid matrix has negatively (or positively) charged groups. In the mobile phase, amino acids with a net positive (or negative) charge move through the matrix more slowly than those with a net opposite charge. So the two types of amino acids can be separated as two distinct bands.

2.3   ACETIC ACID

Acetic acid is used in the production of the polymer, polyethylene terephthalate (PET), which is used in the manufacture of soft drink bottles and fabrics; cellulose acetate, mainly for photographic film; and polyvinyl acetate for wood glue as well as many synthetic fibers and fabrics. Acetic acid is produced both synthetically and by bacterial fermentation. Although, the biological route accounts for only about 10% of world production, but vinegar production is based on this process. About 75% of acetic acid is made for by chemical method by methanol carbonylation.

It is produced aerobically by the genus *Acetobacter*. Genus of *Clostridium* can convert sugars to acetic acid directly, without using ethanol as an intermediate. The product is in dilute aqueous media and contains many organic and inorganic impurities. So the recovery and purification of acetic acid from such a stream is a big challenge. The most important of these challenges is the conversion of the ammonium acetate to acetic acid and ammonium hydroxide. The alkali is then recycled to neutralize the fermentation step. The other hurdles include removal of impurities and water.

The downstream consists of filtration to remove the particles, biomass and solids; electrodialysis to remove and neutralize the salts followed by distillation to recover the acid. Several different designs are available for each of this unit operation. The methods for separating acetic acid from water include: (i) fractional distillation, (ii) azeotropic dehydration distillation, (iii) extractive distillation, (iv) solvent extraction or (v) carbon adsorption.

Out of these, extraction using a hydrocarbon solvent is the most economical method. This step is followed by distillation (Figure 2.4). To avoid product inhibition in the reactor, the acid is continuously removed to keep its level in the fermentor below the toxic value of the organism. The extraction methods that could be adopted are ion-exchange resins, solvent extraction and membrane separation.

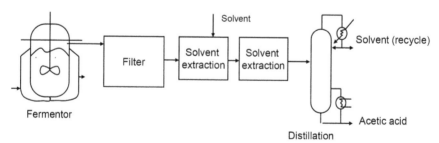

FIGURE 2.4   Schematic of purification steps in the acetic acid process (acetic acid in free acid form).

Filtration is the first step to remove solids after the fermentation. There are three types of filtration namely, depth, screen and surface. Depth filter is made up of compressed matted fibers to form a matrix, which retains particles by entrapment. Screen filter is a porous plate, which retains particles larger than the pore size. Surface filter is made up of multiple layers of media. When the liquid passes through the filter, particles larger than the spaces within the filter matrix are retained. Depth filters are usually used as prefilters because they are economical and can remove ~ 98% of suspended solids and protect downstream from fouling or clogging. Surface filters remove > 99.9% of suspended solids and may be used as prefilters. Microporous membrane filters are placed in the end to remove the last traces.

Electrodialysis (ED) is used to desalt or split water. Bipolar membranes can purify and also neutralize or convert the acetic acid salt back to the corresponding acid and alkali. The capital and operating cost of ED is expensive for bulk chemicals (such as, this acid). Also, the concentration of divalent ion in bipolar membrane operation should be very low.

Organic acids can be produced from these esters also by pervaporation assisted etherification process. Membranes which have a high affinity for water and ammonia, but a low affinity for organics are used here. The assembly has three membranes, made of polyvinyl alcohol top layer, polyacrylonitride middle layer and thermally stable polyester (130°C). The ammonium acetate solution is evaporated and fed to the cracker/separator. There it is heated to cracking temperature of approximately 120–140°C, where the ammonia, water and the acetic acid that is cracked go to the vapor phase. The pervaporation membrane allows only water and ammonia to permeate, while acetic acid does not pass through the membrane. This approach produces acetic acid of 92% purity, which is further purified to 98% using normal distillation columns.

Some of the companies, which manufacture acetic acid are Celanese, BP Chemicals, Millennium Chemicals, Sterling Chemicals, Samsung, Eastman and Svensk Etanolkemi.

2.4 BUTANEDIOL

2,3-Butanediol is a fuel additive. It is used in the manufacture of printing inks, perfumes, fumigants, moistening and softening agents, explosives and plasticizers. It is also used as a carrier in pharmaceuticals. It is produced with *Bacillus polymyxa* or *Klebsiela pneumoniae*. The latter is preferred since it has the ability to use all the sugars (hexoses, pentoses, and certain disaccharides) and uronic acid. The fermentation also generates small amounts of ethanol and the organism is strongly inhibited by it as well as butanediol. Different downstream purification approaches are possible for recovering it from the broth.

Butanediol has a high boiling point so it is not separated in a distillation column but is extracted with a solvent. Several extraction solvents including n-decanol, dibutyl-phthalate, propylene glycol and oleyl alcohol have been tested. But solvents with high extraction capacities are toxic to the cells. Acetone does not affect cell growth and product formation, so *in situ* removal of butanediol using acetone appears attractive. The solvent can be used either with the fermentation medium so that the butanediol gets partitioned into this solvent layer or use an external extraction column in tandem with the fermentor, where the product is removed and the medium is recycled back in to the fermentor. Direct contact between the organic phase and the cells are avoided by means of membranes, which allow continuous processing with continuous product removal. This approach also reduces cell damage and emulsification.

The medium is separated from the cells by a microfiltration membrane and the cell-free medium is contacted with n-decanol in a four-stage, counter–current extraction which prevents direct contact between the solvent and the micro-organisms (Figure 2.5). Butanediol can also be extracted with oleyl alcohol in a Karr extraction column.

Butanediol can also be separated by 'salting out' method using potassium carbonate (94% recovery with 53% out of salt). Precleaning by flocculation is recommended if molasses is used as the carbon source because otherwise the salt cannot be recycled due to strong coloration and contamination. By precleaning 96% recovery is obtained (Figure 2.6).

Companies which manufacture Butanediol include Spectrum chemical manufacturing group, Gardena, CA, USA; Acros Organics; Paragos, Alt Fechenheim, Germany; Matrix, Switzerland; Shangai Rosen Chemicals, Shangai, China; Wintersun chemical, Ontario, CA, USA.

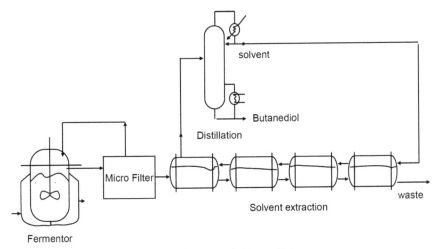

FIGURE 2.5   Schematic of purification steps in Butanediol process.

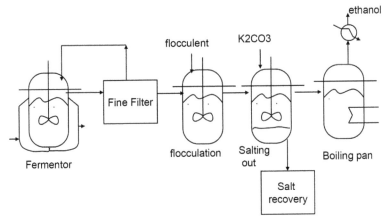

FIGURE 2.6   Schematic of purification steps in Butanediol process (salting out approach).

2.5   STREPTOMYCIN

Streptomycin is the first class of antibiotic drugs called amino glycosides to be discovered, and is the first remedy for tuberculosis. It is produced by the actiniobacterium, *Streptomyces griseus*. The downstream process for this drug is very crucial since it should yield the product meeting the clinical standards with respect to potency, purity and free from toxic contaminant. Manufacturers of streptomycin include Merck & Co Inc., Ranbaxy Laboratories Limited, Pfizer Inc., GlaxoSmithKline and Novartis.

Filtration (such as, leaf or rotary drum) is the first step in the downstream processing of streptomycin to remove dead biomass and cell debris. The isolation of the drug can be done by three different methods, namely, (i) absorption on carbon using non-ionic or activated carbon, (ii) extraction with butyl alcohol, and (iii) dialysis. The active material can be eluted from the carbon with hydrochloric acid with 95% ethanol and then precipitated directly from the solution with ether.

The downs stream process that is commercially practiced consists of (i) clarification of the culture filtrate at a pH of 2 with 0.5% carbon, (ii) extraction of the antibiotics at a pH of 7 with 2% carbon, (iii) washing the carbon successively with water, neutral ethanol, and neutral methanol, (iv) multiple extractions of the carbon with 0.1 N methanolic hydrogen chloride, (v) precipitation of crude streptomycin chloride with ether, and (vi) final purification by reprecipitation from methanol with ether. If methanol solution contains too much water then the product will be a sticky gum.

Alkaline alumina removes streptomycin from neutral aqueous solutions, but elution of the drug later with aqueous acid is slow and incomplete. Acid washed alumina does not remove streptomycin from aqueous methanol.

2.6   BIODIESEL

Biodiesel is a fuel for diesel engines produced from vegetable oils. It is made chemically by reacting a vegetable oil with methanol in the presence of a catalyst, such as, sodium or potassium hydroxide. It produces methyl esters, which is known as biodiesel. Because its primary feedstock is a vegetable oil, it is considered renewable. Biodiesel is considered to contribute much less to global warming than fossil fuels. Diesel engines operated on Biodiesel have lower emissions of carbon monoxide, unburned hydrocarbons, particulate matter, sulfur oxides and other toxics than petroleum-based diesel fuel.

Alcohol, catalyst, and oil are combined in a reactor and agitated for approximately 1 h at 60°C (Figure 2.7). Smaller plants are operated in batch mode, while larger plants (~ million liters/year) use continuous

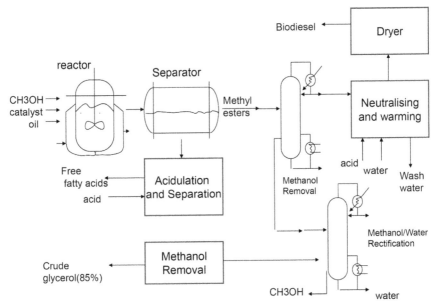

FIGURE 2.7   Flow sheet for the manufacture of biodiesel.

stirred tank reactors (CSTR) or tubular reactors. The reaction is carried out in two steps. In this first reactor, approximately 80% of the alcohol and catalyst are added to the oil. The output stream from this reactor goes through a glycerol removal step before entering a second CSTR. Since the reaction is reversible, removal of glycerol ensures that the reaction shifts to the right of the equilibrium. The remaining alcohol and catalyst are added in the second reactor. Due to the low solubility of glycerol in the esters, separation generally occurs quickly. There are three different approaches by which the ester and glycerol phases can be separated, namely, (i) with the help of a settling tank, (ii) a centrifuge or, (iii) a hydrocyclone.

Decantation is based on the density difference between the two phases and the residence time can vary from 1 to 8 h. The temperature in the decanter affects, (i) the solubility of the alcohol, and (ii) viscosity of the liquids. High temperature in the decanter can cause left over alcohol to flash, while too low a temperature increases the viscosity, which will slow the coalescence rate in the system. While the maximum separation force in the decanter is 1G, higher Gs are created in the centrifuge. The disadvantage of the centrifuge is its initial capital cost, and the need for maintenance. But still they are widely used in the food processing and biodiesel industries.

A liquid-liquid hydrocyclone is a conical vessel where the liquid enters tangentially into the cyclone gets accelerated, similar to the effect in a centrifuge (Figure 2.8). The denser fluid is forced towards the wall and downward, while the lighter fluid is forced to the center and upward. They are extensively used in oil-water separation. The differences in densities determine the separation efficiency while the differences in viscosities determine the resistance to separation. The rapid reduction of pressure in the device leads to flashing of volatile such as, alcohols creating unsafe conditions. Excess methanol should be removed from the system before the mixture is introduced into the hydrocyclone.

Removal of water and other/components can be achieved by using, (i) distillation, (ii) adsorption using carbon or anionic/cationic resins or, (iii) falling film evaporator. A falling film evaporator is a tall empty column operated under vacuum (Figure 2.9). The ester trickles down the inside wall of the evaporator which is kept hot. A thin liquid film is formed in the evaporator. Water and other low boiling components evaporate. Since the liquid is flowing down due to gravity, the contact between the liquid and the hot surface is minimal. So this type of evaporation is ideal for temperature sensitive materials and food products.

There are three side streams that must be treated and they are, (i) excess methanol that needs to be recycled, (ii) glycerol co-product, and (iii) wastewater stream from the process. These various treatment processes will add to the overall operating cost. An excess of methanol is added initially in the reactor to push the equilibrium of the

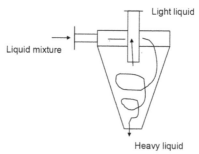

FIGURE 2.8   Liquid–liquid hydrocyclone.

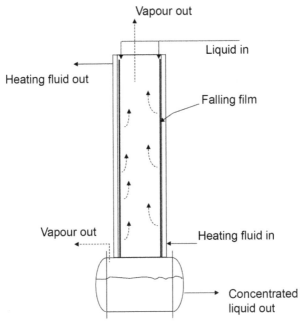

FIGURE 2.9   Falling film evaporator.

transesterification reaction to the right. The recovery of the un-used methanol decreases the raw material cost. Glycerol sales can once again decrease the cost of the final product.

The recovered glycerol will contain residual alcohol, catalyst residue, carry over fat/oil and some esters. It may also contain phosphatides, sulfur compounds, proteins, aldehydes, ketones and insolubles (dirt, minerals, bone or fibers) from raw material oil. Hence, the various steps in the purification of glycerol are:

1. Catalyst is neutralized to salts leading to their precipitation.
2. Soaps produced in the esterification are removed by coagulation and precipitation with aluminum sulfate or ferric chloride.
3. Insoluble or precipitated solids are removed by filtration and or centrifugation
4. Water is removed by evaporation at 70–100°C, where glycerol is less viscous, but still stable.
5. Glycerol is purified using vacuum distillation with steam injection.
6. The glycerol is bleached using activated carbon or clay.

Few companies manufacturing Biodiesel are Gull, a Western Australian based company, Petrobras (the Brazilian national petroleum company), Biox Corporation, Vancouver Biodiesel Co-Op, Biofuel Canada Ltd., Nelson Biodiesel Co-Op, Neste Oil, Finland, Eterindo group, Indonesia, Tesco and Greenergy, UK, Alabama Biodiesel Corporation, Mid-Atlantic Biodiesel, Pacific Biodiesel, Sioux Biochemical, Inc., Michigan Biodiesel, LLC and Johann Haltermann Ltd.

2.7   HUMAN INSULIN

Manufacture of human insulin consists of several steps. Since the product is intracellular, the cells have to be harvested and disrupted to release the product. This is followed by chemical and enzymatic reactions to unfold and refold the protein to achieve the desired activity. Finally the product is purified. Figure 2.10 shows the various processing steps involved in the preparation of human insulin from proinsulin fusion protein. Figure 2.11 gives the breakdown of the cost distribution for the various processes in the flow sheet. 50% of the manufacturing cost is due to raw materials, 17% is equipment dependent and 13% is due to consumables. Reactors and hardware related to fermentation step accounts for 62% of the equipment cost and 29% of the equipment cost is due to the downstream processing units.

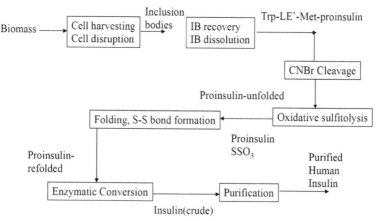

FIGURE 2.10   Downstream of human insulin from proinsulin fusion protein.

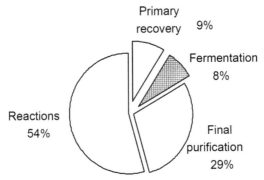

FIGURE 2.11   Cost Distribution in the flow sheet for human insulin production.

KEYWORDS

- **acetic acid**
- **amino acid**
- **biodiesel**
- **butanediol**
- **insulin**
- **streptomycin**

REFERENCES

1. Assman, G., Blasey, G., Gutsche, B., Jeromin, L., Rigal, J., Armengand, R., Cormany, B. Continuous Progress for the production of lower alkayl esters, US Patent No.5, 1996, 514, 820.
2. Carter, H. E., Clark, R. K. Jr., Dickman, S. R., Loo, Y. H., Skell, I. S., Strong, W. A. Isolation and purification of streptomycin, The Journal of Biological Chemistry. July 3, 1945.
3. Jon Van Gerpen, Fuel Processing Technology, Biodiesel Processing and Production. 2005, 86, 1097–1107.
4. Olivio Jose Soccol, Tarlei Arriel Botrel, Hydroclone for pre filtering of irrigation water, Sci. Agric. (Piracicaba, Braz.), Mar./Apr. 2004, 61(2) 134–140.
5. Patents related to acetic acid manufacture are – United States Patent # US2005/0272135 A1 Datta et al.; United States Patent # 4604208, Chaokang Chu et al.; USA Patent #5,972,191, K. N. Mani; USA Patent #US 6,210,455 B1, Lars Olausson et al.; USA Patents # 20050272135, Harry M. Levy et al.

6. Sato, T., Mori, T., Tosa, T., Chibata, I., Furui, M., Yamashita, K., Sumi, A. Engineering analysis of continuous production of L-aspartic acid by immobilized Escherichia coli cells in fixed beds. Biotechnol. Bioeng. 1975, 17, 1797–1804.

7. Stidham, W. D., Seaman, D. W., Danzer, M. F. Method for preparing a lower alkyl ester product from vegetable oil, US patent No.6, 2000, 127, 560.

8. Van Gerpen, J., Shanks, B., Pruszko, R., Clements, D., Knothe, G. Biodiesel Production Technology, Aug 2002–Jan 2004.

9. Vander Brook, M. J., Wick, A. N., Devriers, W. H., Harris, R., Cartland, G. F. Extraction and purification of streptomycin with a note on streptothricin, The Journal of Biological Chemistry, May 27, 1946.

10. Wimmer, T. Process for the production of fatty acid esters of lower alcohols, US Patent No.5, 1995, 399, 731.

11. Yoshiharu Izumi, Ichiro Chibata, Tamio Itoh, Production and Utilization of Amino Acids-review, Angewandte Chemie International Edition in English, 17(3), 176–183.

PROBLEMS

1. Draw two flow sheets for manufacturing an enzyme, which is produced intracellularly by a bacterium. Use different downstream steps in each flow sheet. Product purity desired is 80%. Discuss the advantages and disadvantages of each flow sheet.

2. Draw a flow sheet for producing an antimicrobial peptide from a bacterium as an extracellular product.

3. How can the flow sheet be different if the peptide is thermally stable?

4. Two different downstream flow sheets for butanediol manufacture are shown here. Draw flow sheets two other approaches.

CHAPTER 3

SIZE REDUCTION, BACTERIAL CELL BREAKAGE AND RECOVERY OF INTRACELLULAR MATERIAL

CONTENTS

Biological products can be extracellular (e.g., alcohols, acids, amino acids, antibiotics, enzymes), intracellular (e.g., recombinant DNA products) or periplasmic (e.g., recombinant DNA products). If the bioproduct including enzymes or metabolites remains inside the micro-organism then it has to be broken down to recover the products of interest. Also in many cases the amount of extracellular product secreted by the micro-organism when compared to the amount stored inside may be much less. Intracellular products including recombinant proteins are produced as inclusion body. Then it is prudent to destroy the cell wall to recover the entire product. Generally, intracellular product recovery is expensive. Several additional unit operations are required as shown in Figure 3.1 to recover intracelluar products.

Several mechanical methods are adopted for reducing size of particles. These methods may not be relevant in bioprocess applications, but are widely used in chemical process industries. These methods are useful in the upstream in raw material preparation as well as in the final product polishing stage.

3.1   BACTERIAL CELL BREAKAGE

Gram positive bacteria are 0.5 to 2 μm in size, and have cell walls of 0.02–0.04 μm in thickness. The cell wall is made of peptidoglycan, poly-saccharide and teichoic acid. Gram negative bacteria are 0.5 to 1 μm in size, and the peptidoglycan layer is thin. It has a periplasmic space and has a thick lipopolysaccharide layer. They are mechanically less robust and chemically more resistant than the other bacteria. Yeast (2–20 μm in size, and is spherical or ellipsoid in shape) and molds (bigger than the former and are filamentous) have very thick cell wall. Cell walls in these organisms are mainly composed of polysaccharides including glucans, mannans and chitins. Plasma membranes are mainly made up of phos-pholipids. Animal cells do not have cell walls, are very fragile, microns in size and are spherical or ellipsoid in shape. Plant cells are bigger, have thick and robust cell walls mainly composed of cellulose. They are difficult to disrupt. Cultured plant cells are less robust than real plant cells. Animal cells can be ruptured in a cell disruptor within one pass at

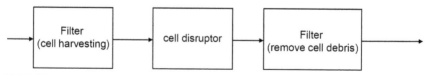

FIGURE 3.1   Unit operations for recovery of intracellular product.

a pressure of 2000 psi; insect blood cells require 15,000 psi; yeast cells require up to 20,000 psi; and plant tissue 40,000 psi.

The breaking of the cell includes cell disruption, lysis, permeabilization, extraction or physical, mechanical, chemical or biochemical method or a combination of these. Liberation of large amount of nucleic acids during this process leads to increase in the viscosity of the broth (generally the fluid become non-Newtonian). Heat shock treatment, which involves heating the contents rapidly followed by holding it to that temperature could denture the DNA/RNA; thereby, preventing the increase in viscosity (Subramanian, 2007).

Several different techniques are adopted to recover the intracellular products and they include the use of (i) high-pressure homogenization, (ii) bead mill, (iii) osmotic shock, (iv) thermal, (v) enzymes, (vi) chemical detergents, solvents including (toluene), urea, antibiotics and lytic, (vii) mechanical methods include ultra-sonication, freeze-thaw cycling. Each technique has its advantages and disadvantages.

3.2   MECHANICAL DISRUPTORS

Cell disruptor and high-pressure homogenizer operate at high pressures (~2500 bars). Both have a positive displacement pump. Homogenizer pressurizes the sample slurry in a chamber and then it releases the contents into another chamber of lower pressure through a valve, which leads to cell breakage. Cell disruptor uses a hydraulic force to accelerate the sample to high pressure and forces it through an orifice. The sample hits the disruption head and enters a second chamber, which is at a lower pressure. Cell breakage is accomplished because of (i) impingement on the valve seat, (ii) high turbulence and shear, (iii) compression produced in the minute gap of the valve opening, and (iv) sudden pressure drop upon release in the low-pressure chamber (Figure 3.2).

Enzymes and proteins are released into the extracellular space (during disruption) at various rates depending upon their location inside the cell. Proteins located in the periplasm are released faster whereas proteins within the cellular components are released at a slower rate. Unbound intracellular proteins are released in a single pass. Membrane bound

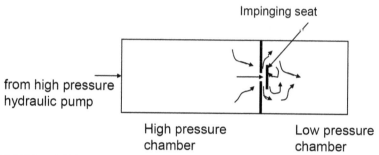

FIGURE 3.2   Cell disruptor.

enzymes or proteins may require several passes through the disruptor. The rate of cell disruption is proportional to

$$d[C]/dt \; \alpha \; v^{1/3} \; \text{and} \tag{3.1}$$

$$v \; \alpha \; \Delta P \tag{3.2}$$

where v = velocity of the slurry flowing C = concentration and ΔP pressure applied.

The release of proteins in the cell disruption process is a first order process and can be represented in an exponential fashion as,

$$[C] = [C]_o (1 - e^{-kN}) \tag{3.3}$$

where, $[C]_o$ – maximum amount of soluble protein; [C] – amount of soluble protein released as a function of time, t; k – constant k = k'P$^{2.9}$ for *Saccharomyces cerevesiae* and k = k'P$^{2.2}$ for *Escherichia coli*; N – number of passes through the homogenizer; P – operating pressure.

The main problem in this technique is the increase in the temperature of the contents, which can lead to denaturation of the proteins that are collected in the medium (Figure 3.3). This denaturation process could be following an Arrhenius type of behavior as shown below:

$$[C]_f = [C] \, e^{-Ed/RT} \tag{3.4}$$

where, $[C]_f$ – amount of active protein finally recovered; E_d – activation energy for deactivation of the protein; T – operating temperature.

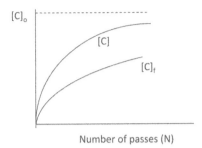

FIGURE 3.3   Intracellular release process (lower release ($[C]_f$) is observed due to deactivation of the protein due to heat generated).

The temperature rise is generally of the order of ~1.5°C/1000 psi. Cell disruption will also release degrading enzymes including proteases, which may cause loss of activity by degrading the protein of our interest. Sudden increase in viscosity after release of contents is another problem that hampers the purification step, which also needs to be addressed during downstream protein recovery.

The French Press is used for disintegrating chloroplast, blood cells, unicellular organisms, animal tissue and other biological solids. This press disrupts the cellular walls of a sample leaving the cell nucleus intact. This unit produces high pressure and rapid decompression to yield cell nucleus. The damage caused by this method is minimal when compared to other mechanical methods. A motor-driven piston inside a steel cylinder develops pressure of the order of 40,000 psi. Pressurized sample suspensions are bled through a needle valve to achieve decompression. The output rate is of the order of about 1 mL/min.

Bead mill consists of a long slowly revolving tube containing ceramic, alumina, steel or glass beads (Figure 3.4). The beads impinge on the cells and cause the breakage of the cell wall and release their contents. The mill

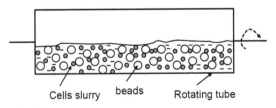

FIGURE 3.4   Bead mill.

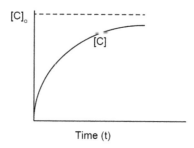

FIGURE 3.5   Intracellular release from a bead mill following a first order kinetics.

rotates at 1500–2250 rpm and the bead diameter is 0.2 to 1 mm. Cell concentration is of the order of 30–60% wet solids. This process is a batch operation. After the cell breakage the contents of the mill are emptied, the cell debris and the intracellular material is separated from the beads, it is washed and, charged again with fresh cell slurry suspension. The separation of the products from the slurry may be achieved by extraction. Strong and hard fibers are also used in the mill, which help in the disintegration of the cells. When rupturing the cells, the objectives are to (i) achieve maximum cell disruption to obtain complete release of cellular contents, (ii) avoid protein denaturation or enzyme deactivation, and (iii) avoid excessive temperature rise which may lead to degradation of products. The release process once again follows a first order kinetics as shown in Eq. (3.5) and (Figure 3.5).

$$[C] = [C]_o (1 - e^{-kt}) \tag{3.5}$$

where k – first order rate constant, and may depend on the hardware. It has to be determined from experimental data and cannot be determined theoretically. This technique is good for cells with tough cell walls, including yeast, spores, and micro algae and also for large-scale operations.

3.3   OSMOTIC SHOCK

Cells exposed to rapid changes in external osmolarity can be damaged or injured. This operation is conducted by first allowing the cells to equilibrate to internal and external osmotic pressure in a high sucrose medium, and then rapidly diluting the external medium. The resulting immediate overpressure of the cytosol damages the cell membrane. Enzymes released by this method are periplasmic or located near the surface of the cell. If the external osmotic pressure changes slowly then the cells will be able to adapt to these changes without undergoing any damage. The osmotic pressure of a fluid containing a salt is given by an approximate equation.

$$\text{Osmotic pressure } \pi = \sum RTC_i \tag{3.6}$$

where C_i – molar concentration of species i. This relation is a simplified version of Van't Hoff's equation.

3.4 ULTRASOUND

Ultrasound with greater than about 18 kHz results in cell lysis, leading to the release of its contents. High frequency produces cavitation and micro-bubbles. The collapse of bubbles converts sonic energy into mechanical energy in the form of shock waves equivalent to 300 MPa pressure. This energy imparts motion to cells, which disintegrate when their kinetic energy exceeds the wall strength. Another factor for the damage of cells is the generation of microstreams. This high velocity gradient causes high shear stress. Cooling is essential in this process. Free radicals, singlet oxygen and hydrogen peroxide are produced during ultrasonication. They will oxidize the metabolites and other products. Use of radical scavengers (e.g., N_2O), reduces the inactivation of the enzymes and proteins. Because of high costs, it is only used in laboratory scale and not for commercial applications. This technique is useful for less resistant cell walls including bacterial and fungal cells.

3.5 PERMEABILIZATION WITH CHEMICALS AND ENZYMES

Many chemicals can permeabilize the outer cell wall that can lead to leakage of intracellular components from micro-organisms. Solvents including toluene, ether, phenylethyl alcohol, DMSO, benzene, methanol, and chloroform create pores in the cell membrane. Chemical permeabilization is performed with antibiotics, thionins, surfactants (Triton, Brij, Duponal) chaotropic agents, and chelates. EDTA (chelating agent) is used for permeabilization of gram negative micro-organisms. It binds to divalent cations namely Ca^{2+} and Mg^{2+}, which stabilizes the structure of outer membranes, by bonding the lipopolysaccharides to each other. Once these cations are removed by EDTA, the lipopolysaccharides are disturbed resulting in increased permeability.

Chaotropic agents, including urea and guanidine facilitate the solubilization of hydrophobic compounds into aqueous solutions. They accomplish this by (i) disrupting the hydrophilic environment of water, or (ii) weakening the hydrophobic interactions between the solute molecules. For example detergents, such as, Triton X-100, in combination with chaotropic agents, such as, guanidine HCl, release membrane-bound enzymes. Major problem in this approach is that these chemicals are costly and are difficult to remove and recover later from the final product mixture.

Enzymes can be employed to permeabilize cells, but this method is limited in releasing periplasmic or surface enzymes only. Some of the enzymes used for permeabilization include $\beta(1–6)$ and $\beta(1–3)$ glucanases, proteases, and mannase. Initially EDTA is used to destabilize the outer membrane of gram negative cells, making the peptidoglycan layer accessible to the enzyme. Later the enzyme can disrupt the cell wall. Lysozyme, from hen egg-white, is the only lytic enzyme available on a commercial scale and is used to lyse gram positive bacteria. The main drawbacks of employing enzymes for recovering intracellular products in large-scale operations are (i) cost and (ii) the difficulty in later removing the lytic enzyme from the product.

Proteins, such as, protamine, or cationic polysaccharide, chitosan, can permeabilize yeast cells. Mammalian cells can be permeabilized by natural substances including streptolysin or viruses. Electrical discharge can also permeabilize mammalian cells.

3.6   SCALE-UP

The method selected for large-scale cell disruption will be different in every case, but will depend on several factors including (i) susceptibility of the cells to disruption, (ii) product stability, (iii) ease of extraction from cell debris, (iv) batch time, and (v) cost.

Factors that have an adverse effect during the cell disruption process are (i) heat, (ii) shear, (iii) presence of proteases, (iv) particle size, (v) DNA and RNA, (vi) presence of chemicals, (vii) foaming and (viii) heavy-metal toxicity. Homogenizers and bead mills are used in industrial scale, while sonication is restricted to lab scale applications only. Hence, it can be deduced that cost plays a very important role in selecting a technique for large-scale applications.

3.7   REDUCTION OF PARTICLE SIZE

This unit operation involves reducing the size of a solid substance from coarse to a powder. It is also known as comminution, grinding and milling. Even cutting is a size reduction process.

Raw materials for a chemical process may vary in their size, shape, brittleness and toughness. The product required may vary from a coarse powder to micron size. Different types of size reduction machinery are used in the industry to suit the specific process requirements.

Objectives of Size Reduction:

- Reducing the size of a solid drug is performed in order to increase its surface area, which can help in its rapid dissolution and absorption or can be mixed and compressed uniformly with other solid materials during tablet manufacturing.
- Before leaching the size of solid materials is reduced since increase in surface area allows the solvent to penetrate easily resulting in quick extraction of the active ingredients.
- Good mixing of several solid ingredients can be achieved if they are all of uniform size.
- The stability and physical appearance of emulsions, ointments, pastes and creams is improved by reducing the particle size.

Several factors related to the solid affect the size reduction process and they are:

Hardness: It is a surface property and is measured by Moh's Scale. A diamond has a hardness of 10 and chalk has a value of 1. Materials that have values above 7 are known as hard materials and below 4 are called soft materials. The harder the material the more difficult it is to reduce its size.

A very hard but brittle material (such as, glass, ceramics of crystals) will not pose any problem. Iron or copper is grouped under soft brittle materials.

Toughness: Rubber is a soft but tough material. It will give more problems during size reduction than a hard but brittle substance such as, wood. Breaking fibrous drugs are difficult since they are tough and the amount of moisture content affects this process.

Toughness of a material could be reduced by dipping it in liquid nitrogen ($< -100°$ C). This process has several advantages, such as:

1. makes the material brittle;
2. decreases the decomposition of thermally labile materials;
3. reduces loss of volatile materials;
4. prevents oxidation of constituents; and
5. avoids risk of explosion.

This is an expensive approach since identifying material of construction of machinery to handle this unit operation will be expensive. Also lubricants may solidify at this temperature.

Abrasiveness: Abrasiveness is a property of hard materials (minerals). Grinding of abrasive substances will damage the equipment and the final powder may contaminate the grinding mill.

Stickiness: Gums or resins are sticky. Here, the material may adhere to the grinding surfaces, or the downstream units. If heat is produced during this process then more problems are caused. Addition of an inert substance (e.g., kaolin to sulfur) may help to reverse this property of slipperiness. Slippery material lowers the grinding efficiency since they act as lubricants.

Softening temperature: Heat is generated during size reduction due to friction and rubbing, which may cause some substances to soften (waxy substances, such as, stearic acid, or drugs containing oils or fats). The operating temperature of the process must be less than the temperature at which this happens. Cooling the mill is one way of controlling this phenomena.

Moisture content: Amount of moisture in the solid influences a number of properties including hardness, toughness or stickiness. The materials to be broken should be dry or wet. Moisture should be less than 5% if the substance needs to be ground dry or greater than 50% if it is being subjected to wet grinding.

Explosion: Some materials may be prone to explosion when subjected to pressure or may produce dust or gasses, which may explode. Performing the size reduction in inert atmosphere or sprinkling with water are a few techniques to avoid these problems.

The change in energy (dE) required to produce a change of dL in the particle of size, L, is represented as

$$dE/dL = KL_n \tag{3.7}$$

where K and n, are constants.

If the energy required to reduce the size of a material is directly proportional to the size reduction ratio dL/L. Then n = −1. So

$$dE/dL = K_K f_c L^{-1} \tag{3.8}$$

where K_K is called Kick's constant and f_c is the crushing strength of the material. If Eq. (3.8) is integrated then energy required to reduce the size from L_1 to L_2 will be,

$$E = K_K f_c \ln(L_1/L_2) \tag{3.9}$$

Equation (3.9) is the Kick's Law. The specific energy required to crush a material from 100 cm to 50 cm, is the same as the energy required to crush the same material from 2 mm to 1 mm.

Rittinger's law assumes that the energy required for size reduction is directly proportional to the change in surface area, that is, n = − 2, then

$$dE/dL = K_R f_c L^{-2} \tag{3.10}$$

$$E = K_R f_c (1/L_2 - 1/L_1) \tag{3.11}$$

where K_R is a constant.

The energy required to reduce particles from 10 cm to 5 cm would be the same as that required to reduce, the same from 5 mm to 4.7 mm. Kick's law can be used if grinding of coarse particles in which the increase in surface area per unit mass is relatively small. Whereas, Rittinger's law is applicable for the size reduction of fine powders in which large areas of new surface are created.

3.8   CLASSIFICATION OF SIZE REDUCTION EQUIPMENT

Size reduction equipment make use of three physical principles namely (i) application of continuous pressure or crushing force, (ii) impact, leading to breaking of brittle material, and (iii) shear forces such as, grinding or abrasion. Other approaches include cutting, slicing, dicing – these operations are required in vegetable and plant processing. Material must be dry to avoid accumulation, jamming and agglomeration in the mill. Materials including camphor and spermaceti tend to cohere and so need wetting with alcohol before size reduction.

Crushers make use of compressive forces to achieve particle size reduction. Jaw and gyratory crushers are two types of crushers (see Figure 3.6). In a jaw crusher, the material is crushed in between two heavy jaws, one fixed and the other reciprocating. The gyrator crusher consists of a conical casing, inside which a head shaped as an inverted cone rotates. The material is trapped between these and crushed. Grinders combine shear and impact with compressive forces.

Roller crushers (Figure 3.7) consist of two horizontal heavy cylinders, mounted parallel and close to each other. They rotate in opposite directions and the material is trapped and hence crushed. The rollers rotate at the same speed. They are used in the sugar cane industry to crush the cane.

Cutting equipment consists of rotating knives, which cut the materials.

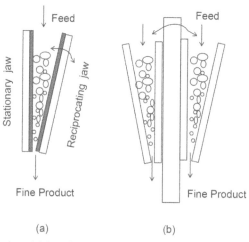

FIGURE 3.6   Crushers (a) jaw (b) gyratory.

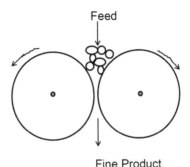

Fine Product

FIGURE 3.7   Roller crusher.

Hammer Mills: Swinging hammers are attached to a rotor, which rotates at a high speed inside a tubular casing Figure 3.8a. The material is crushed and pulverized. The fines pass through a screen located at the bottom and leave the mill. They are separated in sieves. Both brittle and fibrous materials can be handled by these mills. The hammer mills are used in pharmaceutical industry for grinding dry materials, wet filter cakes, ointments and slurries, and in metal processing industries for breaking rocks containing raw metal ores.

Heat is generated due to impact so this is not useful for thermally labile or soft materials. So cooling of the mill may be necessary. The mill may be choked, if feed is not controlled or if it contains fibers. Objects such as, stones or metal in the feed may damage the hammer heads. Magnets may

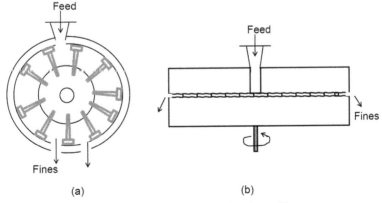

(a)   (b)

FIGURE 3.8   Grinding operation (a) hammer mill (b) plate mill.

be used to remove iron. The hammer mill is capable of producing intermediate sized material, which may have to be ground further.

Plate Mills: The material is fed between two circular plates, one fixed and the other rotating Figure 3.8b. The feed is sheared and crushed near the edge of the plates. These plates are either mounted horizontally or vertically.

Roller Mills: Roller mills are similar to roller crushers, but they are used to finely grind the material. They have smooth fluted rolls, and they rotate at different speeds unlike the crushers. The size reduction of solids is achieved by attrition. They can grind materials that are in suspension, paste, or ointment. Two or three rolls, generally in metal, are mounted horizontally with a very small gap between them. The material is sheared as it passes through the gap.

Colloid Mill: It is used for milling, dispersing, homogenizing and breaking down of agglomerates by shearing action. It is used in the manufacture of food pastes, emulsions, coatings, ointments, creams, pulps, grease, etc. The material is fed as a slurry (Figure 3.9) so that it passes between a rotor and a stator. Combinations of shear and hydraulic forces break the material. The material can be recirculated for a second pass until the desired size is achieved. Cooling and heating jackets are provided. Rotational speed of the rotor varies from 3,000–20,000 rpm. The spacing between the rotor and stator varies from 0.003 to 0.02 cms.

Ball Mill: Ball mill is used to grind materials including ores, chemicals, ceramics and paints. A long cylindrical tube rotates around a horizontal

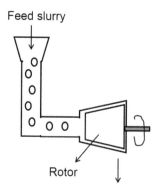

FIGURE 3.9   Colloid mill.

axis partially filled with balls (similar to Figure 3.4). The balls are made of ceramic, flint or stainless steel. As the tube rotates slowly the balls collide with the material and crush and grind the particles to smaller size. They can operate in batch or continuous, and in the latter the material is fed at one end and discharged at the other. Soft or sticky materials cause problems by caking inside the mill. Inflammable material may lead to explosion due to impact. Heat generation is a problem.

Fluid Energy Mill: It consists of a closed loop of pipe, of 20 to 200 mm in diameter (Figure 3.10). Air or fluid is injected at high pressure through nozzles at the bottom of the loop. The circulation of air causes high turbulence and velocity. Solids are introduced into the air stream and they break due to turbulence, impact and attrition. The particle size of the feed to the mill must be of the order of 150 microns to yield a product of size less than 5 microns. After the size reduction operation the particles are separated downstream based on their size by sieving, sedimentation or cyclone separation. Large particles are again recycled to the size reduction equipment.

3.8.1   CYCLONE SEPARATOR

The cyclone separator is a cylindrical vessel with a conical bottom (Figure 3.11). The slurry (solids suspended in gas or liquid) is introduced tangentially at high velocity at the top. The fluid is removed from a central

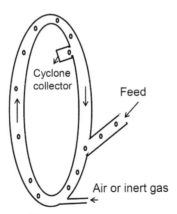

FIGURE 3.10   Fluid energy mill.

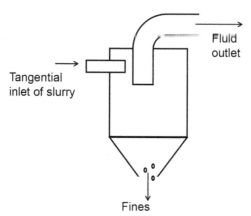

FIGURE 3.11   Cyclone separator.

outlet at the top. The tangential flow within the cyclone causes the particles to be thrown out to the walls, and slide down to the bottom discharge outlet. The cyclone can be used to separate solids from air (in pollution control system) or coarser ones from finer ones. The fines may get carried away at the top by the air.

KEYWORDS

- **cell breakage**
- **cyclone separator**
- **FRENCH Press**
- **high pressure homogeneizer**
- **osmotic shock**
- **size reduction**

REFERENCES

1. Asenjo, J. A. (Ed), Separation process in biotechnology, CRC Press, ISBN 0824782704, 9780824782702, 1990.
2. Badger, W. L. Introduction to Chemical Engineering.

3. Brookman, J. Mechanism of cell Integration in High Pressure Homogenizer, Biotechnol. Bioeng, 1974, 16, 371–383.

4. Bulock, J., Christiansen, B. Basic biotechnology, Academic Press, 1989.

5. Lachman, L. The Theory and Practice of Industrial Pharmacy.

6. Martin Chaplin, Christopher Bucke, 'Enzyme Technology,' Cambridge University Press, 1990.

7. Naglak, T. J., Hettwer, D. J., Wang, H. Y. Chemical permeabilization of cells for intracellular product release in Separation Processes in Biotechnology, Marcel Dekker Inc., NY 1990.

8. Remington – The Science and Practice of Pharmacy, 20th Edition, Volume 1.

9. Subramanian, G. (Ed), Bioseparation and Bioprocessing, A Handbook, Wiley VCH, Weinheim, 2007 (ISBN 978-3-527-31585-7).

PROBLEMS

1. Assume the protein deactivation kinetics to follow a first order kinetics with activation energy of 30 Kcals/mol. If the protein release rate from *E. Coli* in a bead mill is a first order kinetics with the release constant is equal to 0.1/min and the maximum concentration of intracellular protein is 10 mmol, estimate the maximum amount of protein that can be recovered using this operation.

2. If a high-pressure homogenizer operating at 200 bars is used to recover the same protein, how many passes are required to recover the maximum amount of protein. Assume k' = 0.1.

3. For a disruption of bacteria, a homogenizer was found to give a yield of 50% when operated at 50 MPa, and 90% when operated at 120 MPa. Determine the number of passes required to achieve 90% yield at 50 MPa.

4. For a 1% increase in the pressure in a homogenizer what will be the increase in temperature.

5. Develop a mathematical model for the breakage of cell wall when subjected to enzymatic lysis. Assume (a) first order kinetics (b) Michaelis Menton relation.

6. For a double-layered cell wall protease and glucanase are used to degrade the outer and inner cell walls respectively. If the former has an optimum pH of 9 and the later has a pH of 6, how could one decease the degradation of product by protease.

7. Develop a mathematical model for the breakage of cell wall when subjected to high-pressure homogenizer. Assume first order kinetics for material release as well protein deactivation.

8. Use Kick's Law and Rittinger's law to determine the specific energy required to crush particles from 10 mm to 4 mm in size. Assume all the constants to be equal 1.

9. At what particle size values will Kick's law predict higher energy than Rittinger's law.

10. If we assume 10% of the energy put in for crushing/grinding leads to heat generation then will more heat will be generated to break particles from 10 to 5 mm, or from 1 to 0.5 mm?

CHAPTER 4

ISOLATION OF INSOLUBLES

CONTENTS

Removal of solids is invariably the first step in the downstream. The dead biomass, cell debris and other insoluble salts from the fermentor broth are removed by one of these techniques. Immobilized biocatalyst is also removed and recycled for reuse. If the desired product is intracellular then the cells have to be harvested for lysis. So in the last two situation, the solid is the desired product, whereas in the first case the solids are collected and disposed. Treatment of liquid effluent also requires the removal of insoluble or suspended solids. There are several different solid removal techniques that are common to both, chemical and biochemical processes.

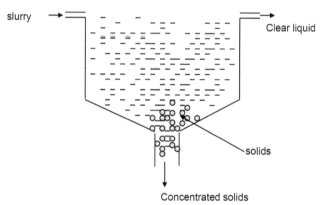

slurry

Clear liquid

solids

Concentrated solids

FIGURE 4.1   Settling tank.

4.1   SETTLING/SEDIMENTATION

This is the easiest of the unit operations and, it needs a large vessel and
sufficient time for the solids to settle down (Figure 4.1). At times, the
solids may not settle down due to electrostatic forces, which may repel
them or their small size. Then one may have to use coagulating agents
or antistatic agents. The settling time depends on the physical properties
of the solid and the fluid; and under dilute conditions the Stoke's law of
settling determines the terminal settling velocity. When the buoyancy on
the particle exactly matches with the drag force, then the particle reaches
a constant (terminal) settling velocity under the force of gravity as shown
below,

$$v = \frac{d^2}{18\mu}(\rho_s - \rho_l)g_c \qquad (4.1)$$

where, v = terminal settling velocity of the solid in a dilute solution. This law
is valid only when Reynolds number, $N_{Re} = \left(v\dfrac{d\rho_l}{\mu} \right) < 1.0$; d = diameter of
the particle (for a non-spherical particle, the ratio of volume to surface area
of the particle can be taken as the characteristic radius); ρ_l = liquid density;
ρ_s = solid density; μ = viscosity of the fluid; g_c = acceleration due to gravity;
N_{Re} = Reynolds number.

If the settling is under the influence of centrifugal force (as it happens in a centrifuge), then the terminal settling velocity for the particle will be

$$v = \frac{d^2}{18\mu}(\rho_s - \rho_l)\omega^2 r \qquad (4.2)$$

where, ω = angular rotation of the centrifuge, rad/sec; r = distance of the solid particle from the center of the axis of the centrifuge; g_c is the driving force in Eq. (4.1) and $\omega^2 r$ in Eq. (4.2).

4.2   FILTRATION

Two main types of filters are:

1.   Surface filter – a solid sieve, which traps the solid particles, with or without the aid of filter cloth (Figure 4.2a). Here, the particulates are captured on a permeable surface (e.g., Büchner funnel, Belt filter, Rotary vacuum-drum filter, Cross flow filters, Screen filter).
2.   Depth filter (Figure 4.3) – a bed of granular material, which retains the solid particles as it passes, for example, sand filter.

The first type allows the solid particles, that is, the residue, to be collected intact. The second type does not permit this and it is predominantly used in water purification systems.

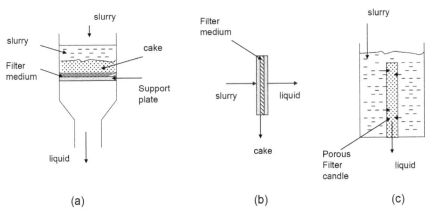

(a)   (b)   (c)

FIGURE 4.2   Filter setup (a) horizontal design, (b) plate and frame design, and (c) candle filter.

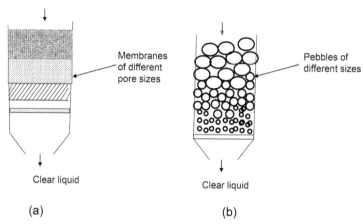

FIGURE 4.3   Depth filter (a) membrane and (b) pebbles.

In the surface filter the slurry is passed through a cloth, metal, polymer, fibers or candle with fine pores. The solids are retained and the liquid
flows through it. In the horizontal design (Figure 4.2), the solids remain
on top of the filter material. The liquid flow is achieved either by applying
pressure from the top or vacuum from the bottom of the unit as shown in
Figure 4.4. Accumulation of solids on the filter media leads to reduction
in the rate of filtration. So separation process has to be stopped and the

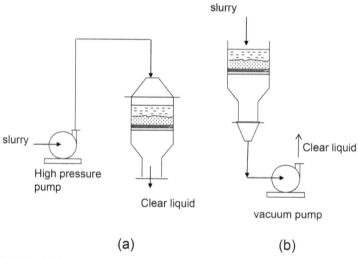

FIGURE 4.4   (a) Pressure or (b) vacuum filter setup.

surface has to be cleaned. The plate and frame design consists of several plates through which the liquid flows and the solids are retained in the filter material (Figure 4.2b). Here, several plates are stacked together to achieve the desired filter area. The candle design consists of a porous candle through which the liquid flows (Figure 4.2c). Metal or ceramic material is used here and they are suitable for hard or abrasive materials.

In cross flow filtration the slurry does not flow perpendicular to the filter medium, but flows parallel to it. This prevents the accumulation of the solids and clogging of the filter media.

Sand filter is a depth filter (Figure 4.3b). Pebbles and sand of various sizes are placed here while liquid passes through this unit. Particles and dust are retained when they come in contact with the filter material. This is suitable if the amount of suspended particles that need to be removed are less, especially in water purification systems. Depth filter can also contain a series of membranes of various pore sizes (Figure 4.3a) and solids of various sizes will be retained at different depths.

The commonly used filter media are:

- filter paper;
- woven material;
- cheese cloth, woven polymer fiber, woven glass fiber);
- non-woven fiber pads;
- sintered and perforated glass;
- sintered and perforated metals;
- ceramics; and
- synthetic membrane (made of polymers).

The smallest particles retained by various materials in various forms and rigidity are listed in Table 4.1. These values are only approximate numbers.

The rate of filtration is proportional to filter area, pressure gradient and bed thickness and is inversely proportional to the viscosity of the fluid. Darcy's law relates these parameters as shown below

$$v = k\, \Delta p / \mu * b_d \qquad\qquad (4.3)$$

where, v = velocity of liquid through the bed of solid; k = constant; Δp = pressure drop across the bed; b_d = bed thickness; μ = viscosity of liquid.

TABLE 4.1   Types of Filter Medium and Smallest Particles They Can Retain

Solid form	Flat, wedge-wire screen	100 μm
	Wire-wound tubes	10
	Stacked discs	5
Metal sheet	Perforated	20
	Sintered woven wire	1
	Unbonded mesh	5
Rigid porous	Ceramics	1
	Sintered metal powders or fiber	1
	Carbon	1
	Sintered plastic powder or fiber	<1
Cartridge	Yarn wound	5
	Bonded granule or fiber	1
	Pleated sheet	<1
Plastic sheet	Perforated	10
	Sintered woven filament	5
Membrane	Ceramic	<0.1
	Metallic	<0.1
	Polymeric	<0.1
Woven	Stable polymeric fiber yarn	5
Non-woven	Dry-laid	10
	Wet-laid (paper)	2
	Wet-laid (sheets)	0.5
Loose	Fibers	1
	Powders	<0.1

This law holds only when

$$N_{Re} = d\,v\,\rho/\mu\,(1-\varepsilon) < 5 \tag{4.4}$$

where, d = particle size or pore diameter in the filter cake; ρ = liquid density; ε = void fraction in the cake; N_{Re} = Reynolds number.

In a batch filtration process, a certain quantity of the slurry is filtered and the filtration unit is cleaned to remove the collected solids. Then the process is restarted again. If a slurry batch of V is to be filtered then

$$v = (1/A) (dV/dt) \qquad (4.5)$$

where, V = total volume of filtrate; t = time; A = filtration area.
Combining equations (4.1) and (4.3) give,

$$dV/dt = k \, A \, \Delta p/\mu^* \, b_d \qquad (4.6)$$

The filter cloth/medium also offers some resistance to flow, which may be generally neglected.

For incompressible cake, the amount of solids in the slurry, the bed height (or cake thickness) and area of the filter will be related as

$$b_d = \rho_o \, V/A \qquad (4.7)$$

where, ρ_o = mass of cake solids per volume of filtrate; Ω_c = specific cake resistance (=1/k).

The equation can be integrated assuming at $t = 0$, $V = 0$, then the time required to filter a volume of liquid V in a batch unit will be

$$t = (V^2 \, \mu \, \Omega_c \, \rho_o)/ (2 \, A^2 \, \Delta p) \qquad (4.8)$$

In the graph between V versus t/V if the line passes through the origin then it can be concluded that the resistance offered by the filer cloth is negligible. Otherwise, the intercept indicates the extent of cake resistance.

When the cake is compressible, then increasing pressure increases cake resistance and hence the filtration rate will drop. Then

$$\Omega_c = \Omega_c{}'(\Delta p)^s \qquad (4.9)$$

where, s = cake compressibility, and is 0 for rigid incompressible cake and 1 for highly compressible cake. The value of s is generally between 0 and 0.8. If s is high then filter aids including celite or sand is added to reduce the compressibility and improve filtration.

4.3   CONTINUOUS ROTARY FILTERS

This consists of a rotating drum with vacuum applied inside, so the solid accumulates on the outer surface as the drum dips into slurry tank while the clear liquid is sucked inside (Figure 4.5). The cake that is formed is washed and finally scrapped. A filtration cycle in a rotary filter consists of three main steps namely, cake formation, cake washing (use of water or solvent to remove impurities) and cake discharge.

The cake formation begins as the rotating drum dips into the slurry. If the resistance of the filter medium R_M is negligible, then the same basic expression (Eq. (4.8)) for filtration can be used,

$$(1/A)\,(dV/dt) = \Delta p/\mu\,R_C \qquad (4.10)$$

As before, this is subject to the initial condition that

$$t = 0,\ V = 0$$

For compressible cake

$$t_f = [\mu\,\Omega_c'\,\rho_o(2\,\Delta p)^{1-S}](V_f/A)^2 \qquad (4.11)$$

where t_f is the cake formation time and V_f is the volume of filtrate collected during that period. This relationship is sometimes written in terms of the overall cycle time, t_c as

$$V_f - \beta\,t_c \qquad (4.12)$$

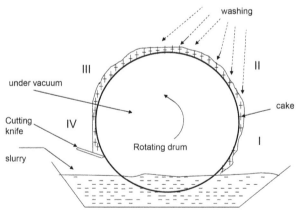

FIGURE 4.5   Rotary drum filter setup. Stage I – cake formation, II – washing, III – drying, IV – cake removal.

where β is fraction of time that the filter is submerged, that is, the fraction of the cycle devoted to cake formation. Combining these equations a relation for filtration flux expressed in terms of cycle time is obtained.

$$(V_f A t_c) = [2 \beta (\Delta p)^{1-S/} \mu \Omega_c' \rho_o t_c]^{1/2} \tag{4.13}$$

The cake formation can be altered by varying either the total cycle time, t_c, or the fraction of the total cycle time devoted to cake formation β. At constant β, the filtration flux is inversely proportional to the square root of cycle time.

The cake may contain some amount of impurities and liquid broth. This is removed by washing the cake with water. The washing has two functions namely, it displaces the broth trapped in pores in the cake and then it allows diffusion of impurity out of the biomass in the cake. The fraction of soluble material remaining after the wash is given by,

Impurity remaining after the wash/originally present $= (1 -e)^n$ (4.14)

where, n is the volume of wash liquid divided by the volume of liquid retained in the cake and e is the washing efficiency.

From Eq. (4.14) one can estimate how much liquid will be needed to wash the cake. The wash liquid will not contain additional solids so the flow of wash liquid will be constant and equal to final instantaneous filtrate rate, t, at the end of cake formation. This rate is

$$(V_w/A) = [2(\Delta p)^{1-S/} \mu \Omega_c' \rho_o t_f]^{1/2} * t_w \tag{4.15}$$

where, V_w is the volume of wash water required and t_w is the time required for the washing. Dividing this equation with Eq. (4.9) one gets

$$t_w/t_f = 2nf \tag{4.16}$$

4.4   CENTRIFUGATION

A centrifuge is a device that separates particles from a slurry according to their size, shape and density due to centrifugal force. Centrifuges are used in industrial scale as well in labs. Centrifuges can be used to remove cell

debris, biomass and other large biomolecules after fermentation. For separating proteins, carbohydrates ultra-high centrifuges are used. Centrifuge
can also be used to separate a heavy phase or a more denser liquid from
a mixture of liquid phases. Sometimes the solid is the desired product
while at other times the liquid is the desired product. Centrifuge is used to
concentrate solid phase by removing the excess liquid phase, remove grit
(degritting), and classify liquid (remove light particles).

The magnitude of the gravitational field obtained from a centrifuge is
measured in terms of the G value:

$$G = 1.12 \times 10^{-3} \, r \, (RPM)^2 \qquad (4.17)$$

where, r = distance from the axis of rotation (m).

Less than 3000 rpm is sufficient to separate mixed liquids; while tiny
particles or those phases with little density differences, the speed should be
between 8000 to 30,000 rpm. Ultra high centrifuges rotate at 100,000 rpm.
Centrifuges can be divided into two major types, stationary and rotating
devices, but both types work on the same common principle. The faster a
particle moves, the greater will be the tendency for it to escape from the
central shaft, and move to the periphery.

In a stationary centrifuge, a fluid (a gas or liquid) consisting of two
or more components is introduced tangentially into a cylindrical or conical chamber at a high speed (Figure 3.11). As the fluid travels inside the
chamber the centrifugal force pushes the heavier substance to the walls of
the chamber and the lighter substance remain at the center. It is used in the
downstream of air filters to capture fine solids that escape at the exhaust
from boilers, chimneys, etc.

Another application of the stationary or tangential flow centrifuge is in the separation of the isotopes of uranium from each other.
Naturally occurring uranium consists of a mixture of uranium-235 and
uranium-238. Uranium is first converted into the gaseous uranium hexafluoride and then injected into this cyclone. The heavier isotope 235
reaches the outer wall and the lighter isotope-238 remains at the center
of the cyclone.

Tubular bowl centrifuge is a vertical unit, mounted (Figure 4.6) with
a rotating bowl with ratio of length to diameter from 5 to 7. Feed slurry is
introduced from the bottom. The solids accumulate at the tube wall, which
needs to be scrapped off from time to time. Generally, it is useful for low

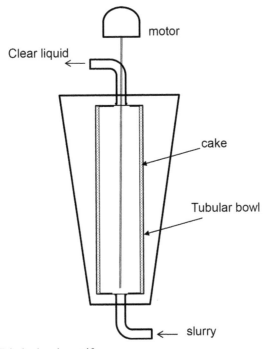

FIGURE 4.6   Tubular bowl centrifuge.

solid concentration. Chamber bowl centrifuge (Figure 4.7) contains number of tubular bowls arranged co-axially. The solids settle onto the outer wall of each chamber.

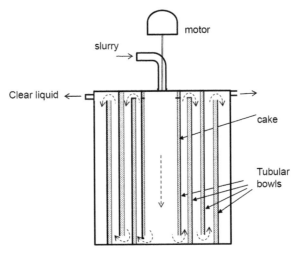

FIGURE 4.7   Chamber bowl centrifuge.

A disc stack centrifuge (Figure 4.8) is a compact design and gives better solid-liquid separation than the tubular bowl centrifuge. The feed enters and is distributed at the bottom of the bowl. The particles are thrown outward when they come in contact with the angled disc stack. The liquid flows up the device along the central region and is discharged from the top.

In centrifugal filter (Figure 4.9) the wall of the drum is porous so the solid is retained inside the filter while the liquid escapes by traveling

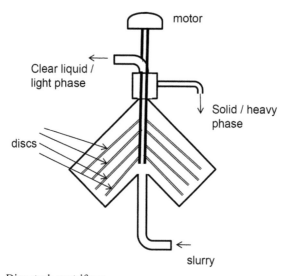

FIGURE 4.8   Disc stack centrifuge.

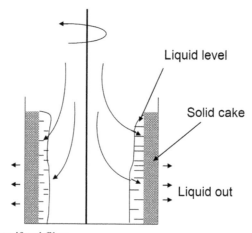

FIGURE 4.9   Centrifugal filter.

through the solid bed. When the bed reaches sufficient thickness, the centrifuge is stopped and the accumulated solid is removed by scrapping. Centrifugal filters are more efficient than conventional filtration for removing the liquid.

The throughput through a centrifuge is dependent on hardware and the physical properties of the solid and the liquid,

Throughput = fn (particle settling velocity, hardware parameters);
Throughput = fn (physical properties of the solid & liquid, hardware).

For a Tubular bowl centrifuge the throughput will be,

$$Q = v_g [2 \pi l R^2 w^2/g] \qquad (4.18)$$

R is the bowl radius, l = length of the tube, ω = the angular velocity, v_g = the terminal settling velocity.

For a Disk Stack centrifuge

$$Q = v_g [(2 \pi n w^2/3 g) (R_0^3 - R_1^3) \cot \theta] \qquad (4.19)$$

where, n = number of disks, R_0 and R_1 are the outer and inner radius of the bowl and θ = angle the disk makes with the vertical axis.

4.4.1 LABORATORY CENTRIFUGES

An ultracentrifuge is generally used in laboratories. Rotation speeds of 30,000–50,000 rpm are achieved in these equipment. It is usually used for separating macromolecules and biological components. A centrifugal filter is a laboratory instrument can be a micro-spin tubes (15–20 mL), with a membrane module as an insert for separation or a chromatography column for absorption and purification.

Density gradient centrifugation involves use of a specific medium that gradually increases in density from top to bottom of the centrifuge tube. Under centrifugal force, particles will move through this medium and stop at a point in which the density of the particle equals the density of the surrounding medium. These techniques can be used for separating a mixture of particles/biomolecules according to their densities.

Examples of gradient materials include,

- Simple sugars – sucrose, sorbitol, glycerol
- Polysaccharides – Ficoll, dextran, glycogen
- Proteins – bovine serum albumin
- Inorganic salts – CsCl, Cs_2SO_4
- Iodinated organic compounds – metrizamide, renographin
- Solvents – DMSO, formamide

4.5   COAGULATION AND FLOCCULATION

Very fine particles may not settle in a settling tank or it may be difficult to filter them. The charge associated with fine particle prevents them from agglomeration. In such situations one need to resort to these techniques for removal of the solids.

- Coagulation is the process through which colloidal particles and very fine solid suspensions are destabilized so that they begin to agglomerate and form large particles. Chemicals, polymers or salts are added to achieve coagulation.
- Flocculation is the process by which destabilized particles conglomerate into larger aggregates (or flocs).

Hydrophilic colloids are formed by organic macromolecules that become hydrated/solvated when they are in the presence of water. These molecules are thermodynamically stable in their solvated form. Agglomeration of hydrophilic colloids can be achieved by adding ions, which compete with the water molecules in the colloids thus resulting in the dehydration of the colloidal particles ("salting out" of the colloid). Hydrophobic colloids are those that do not have an affinity for water.

The presence of a charge, which attracts other ionic species present in the water results in the formation of an electrically charged layer around the colloidal particles. Colloidal dispersions are thermodynamically unstable. If the charge layer is removed they tend to agglomerate spontaneously and can be removed from the water.

If a colloidal particle is electrically charged it attracts ions and other colloidal particles of opposite charge. These ions are tightly bound by electrostatic forces to the colloidal particle forming a charged inner layer known as the 'Stern layer.' This has a thickness of a single hydrated ionic layer. These attached ions are of opposite sign known as electric double layer. Ions of opposite sign to that of the Stern layer attach to the latter and this is known as the diffuse layer.

The zeta potential is defined as the electric potential difference between the shear plane of a colloidal particle and the bulk of the solution. It is an indirect measure of the electrical charge of the colloidal particle. When these particles are in the presence of counter ions they become electrically neutral. This point is called the isoelectric point. The zeta potential at the isoelectric point is zero.

The common coagulants used in water treatment are:

Aluminum salts (alum), Ferric and ferrous salts, Lime, Cationic polymers, Anionic and non-ionic polymers. Polyelectrolytes can be classified according to their origin as: natural, that is, derived from starch products or of biological origin (e.g., alginate, chitosan). synthetic, (e.g., polyamine, sulfonate, etc.).

Cationic polyelectrolytes are used to coagulate colloids that are negatively charged. The mechanisms here are: charge neutralization and bridging of colloidal particles.

After the colloid particles are destabilized, flocculation of these occurs due to two reasons namely, (i) Perikinetic flocculation occurs due to the Brownian motion of the destabilized small colloidal particles. The Brownian motion allows the particles to come close enough for agglomeration to occur. (ii) Orthokinetic flocculation occurs due to the velocity gradients such as, mild agitation which promotes the aggregation of the particles and hence flocculation. Nonionic and anionic polyelectrolytes can be used to destabilize negative colloids. In this case the destabilization mechanism is believed to be due to bridging.

Factors influencing the efficiency of flocculation process include,

- type of coagulant used;
- its dosage, amount and concentration;
- pH of the solution;
- type and dosage of other chemical additives (e.g., polymers);
- sequence of chemical addition and time lag between dosing points;
- intensity and duration of mixing;
- type of mixing device;

- velocity gradients generated;
- retention time;
- flocculator geometry.

4.6   PRETREATMENT OF FERMENTATION BROTH

Fermentation broth contains fine colloidal suspension, which is difficult to remove by filtration, so pretreatment of the broth is carried out before it is sent for filtration. The fines removal is also known as clarification.

Few of the pretreatment methods involve,

1. Heating
2. Coagulation and flocculation
3. pH adjustment. This will selectively denature proteins or allow fines to coagulate or precipitate. Examples include

	pH
Tris and acetic acid	4.5–8.5
Lactic acid	<3.5
Diethanolamine	<9
Sodium carbonate	<10.5
NaOH/KOH	<11
Phosphoric sulfuric acid	2–3

4. Use of filter aids which facilitate liquid flow through the bed by increasing the porosity or adsorption of the fines. Filter aids are inert and incompressible particles of high permeability. Examples include wood pulp, starch powder, cellulose, inactive carbon, diatomaceous earth, rice husk. Their sizes are generally in the range of 2–20 μm.

Disadvantages of filter aid include they are not suitable for large particles and certain molecules may bind irreversibly to filter aids. Examples include aminoglycoside may bind to diatomaceous earth, different filter aids have to be tested before arriving at the best material and concentration and contamination of the solid product by the filter aid.

Filter aids can be used in three ways and they are:

1. Mix it with the slurry and feed to the filtration equipment.

2. Impurities and filter aid are removed together on the top of the cake or

3. As a pre-coat to prevent the filter cloth becoming blocked during the actual filtration.

KEYWORDS

- **centrifugation**
- **coagulation**
- **filtration**
- **flocculation**
- **pretreatment**
- **settling**

REFERENCES

1. Centrifugation: A Guide to Equipment Use and Maintenance, the Biomedical Scientist, Feb 2013, 76–77.
2. Christopher Dickenson, T. Filters and Filtration Handbook, Elsevier, Technology and Engineering; 1997, 1079 pages.
3. Ken Sutherland, A-Z of Filtration and Related Separations, Elsevier, 2005.
4. Read more: Centrifuge – Types of Centrifuges – Uranium, Hexafluoride, Stationary, and Isotope – J Rank Articles, http://science.jrank.org/pages/1338/Centrifuge-Types-centrifuges.html#ixzz3ASaDYLJs

PROBLEMS

1. We want to filter 15,000 L/hr. of a broth containing a biopharmaceutical product using a rotary vacuum filter, which has a cycle time of 50 sec, operating under a vacuum of 20 in. Hg. The pretreated broth forms an incompressible cake and $\mu \, \Omega_c \rho o/\Delta p = 58$ sec/cm^2. We want to wash the cake we expect that the washing efficiency will be 70% and that 1% of the filtrate is retained. (a) Calculate the filtration

area if filtration time per cycle is 10 sec. (b) If the washing time is 2 sec estimate how much impurities are retained in the cake.

2. A laboratory column is used to purify a peptide and we obtain satisfactory results under the following conditions: velocity 30 cm/hr., bed height 15 cm, column diameter 2.5 cm, temperature 25°C, diameter of adsorbent 75 mm, and pressure drop across the bed 3.75 kg/cm². At constant velocity, the pressure drop is inversely proportional to the particle surface area. We want to scale-up the process by increasing the bed volume a thousand-fold and the column diameter tenfold. If the same superficial velocity is to be maintained, what will the pressure drop across the bed be?

3. Derive the equation for terminal settling velocity for a solid moving under acceleration due to gravity.

4. Then much reduction in through put will be observed if the same centrifuge is used to remove particles of half the diameter.

5. If the diameter of a tubular bowl centrifuge is doubled what will be the increase in through put.

6. If the particle size decrease by ¼, what will be the change in setting time. Suggest a few methods for speeding up setting.

7. What will be the change in filtration time (in a normal filter) if the solids is compressible with s = 0.6, when the pressure is 1 bar to 10 bar.

8. In a laboratory filtration it takes 10 min to filter 2 L of slurry and 18 min to filter 3 L of slurry. Does the filter medium offer any resistance?

9. In problem 8 if the filter is doubled what will be change in filtration time.

10. How will the settling velocity of a particle change if the rpm of the centrifuges is increased by 25%?

CHAPTER 5

PRODUCT RECOVERY

CONTENTS

This chapter deals with various product recovery techniques such as, membrane processes, dialysis, electrodialysis, liquid–liquid extraction, adsorption, and precipitation. These techniques help in the recovery of the product from the reaction mixture, but more steps may be required to achieve high purity product.

5.1 MEMBRANE PROCESSES

The membrane separation process is based on the use of semi-permeable membranes It acts as a filter, except in the case of reverse osmosis (RO) and pervaporation that will let the solvent flow through, while it holds suspended solids and other substances. There are various methods to enable solvents to flow through a membrane and they include application of (i) high pressure, (ii) concentration gradient on both sides of the membrane, or (iii) applying an electric potential. Membrane filtration is an alternative technique to simple filtration, flocculation, sedimentation, adsorption, extraction and distillation. The first three techniques deal with the removal of solids from a solution. Membrane filtration is used in food, dairy, desalination, and waste treatment plants.

Membrane filtration can be divided into (i) micro and ultra filtration and (ii) nano filtration and RO. When membrane filtration is used for the removal of larger particles, micro filtration and ultra filtration are used. When salts need to be removed from water, nanofiltration and RO are used. The last two techniques do not work based on flow through pores. The separation takes place by diffusion through the membrane. The pressure that is required to perform these two techniques is much higher than the pressure required for the former, and the productivity is much lower than the former. Membrane filtration systems can be operated as direct flow (dead-end flow or vertical flow) or cross-flow (Figure 5.1). Membranes are designed in the form of tubular membrane and the plate and frame membrane (Figure 5.2). The former is further divided as tubular, capillary and hollow fiber membranes. The last is divided as spiral membrane and pillow-shaped membrane.

In ultrafiltration (UF), high molecular weight compounds such as, proteins, and suspended solids are rejected (retained), while all low molecular

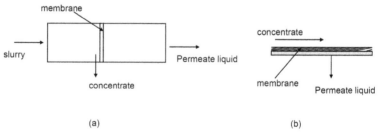

(a)   (b)

FIGURE 5.1   Membrane setup (a) Direct flow (b) Cross flow.

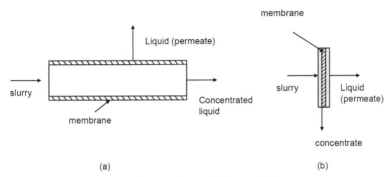

FIGURE 5.2 Designs (a) Tubular Membrane (b) plate and frame.

weight compounds pass through the membrane. Mono- and di-saccharides, salts, amino acids, organics, inorganic acids or sodium hydroxide pass through the membrane. In microfiltration (MF), suspended solids are rejected, while even proteins pass the membrane freely. RO water is the only material that passes through the membrane; and all dissolved and suspended materials are rejected. So in nanofiltration (NF) only ions with more than one negative charge, such as, sulfate or phosphate will be rejected, while single charged ions are passed. It also rejects uncharged molecules, dissolved materials and positively charged ions depending on the size and shape of the molecule. RO and NF membranes are generally made of cellulose acetate (CA).

UF membranes are made up of ceramic, polysulphone (PSO), polyvinylidene difluoride (PVDF) and CA. MF membranes are made up of ceramic, polypropylene, PSO, and PVDF. Generally, the operating pressure is 15–150, 5–35, 1–10 and <2 bar for RO, NF, UF, and MF, respectively. The pore size of RO and NF membranes are < 0.002 μm and that of UF and MF are 0.2–0.02 μm and 4–0.02 μm respectively. Generally, the membranes are about 150 μm thick and thin membranes are of 1 μm thick. Total worldwide consumption of membranes (2001), based on its surface area, is Composite RO membranes: 85%, Composite NF membranes: 3 – 5%, Polysulfone UF and MF membranes: 5–7%.

There are two factors that determine the performance of a membrane filtration process and they are: (i) selectivity, and (ii) productivity. Selectivity is expressed as retention or separation factor and productivity is expressed as flux.

The benefits of using Membrane filtration technology for water purification over the other existing techniques are,

1. It can be performed at low temperature. So it can be used for heat sensitive materials such as, in food and diary industries
2. Energy that is required to pump liquids through the membrane is less when compared to alternative techniques, such as, evaporation, which requires more energy.
3. The process can easily be scaled up. Adding few more membrane modules increases the throughput.

One needs to consider several factors while designing a membrane system and they include (i) capital cost, (ii) operating cost, (iii) plugging of the membranes, (iv) its packing density, (v) stability at various conditions, (vi) performance, and (vii) ease of cleaning. The fouling of a membrane is due to feed water quality, membrane type, membrane raw materials, process design and control. Suspended particles, biofouling and scaling are the three main types of fouling on a membrane. A prefilter can remove suspended particles. Scaling occurs due to the precipitation of dissolved salts. Biofouling occurs due to micro-organisms present in the water, which may settle and start colonizing the surface. There are a number of cleaning techniques for the removal of membrane fouling. These techniques are forward flushing, backward flushing, air flushing and chemical cleaning. Addition of biocide can prevent the formation of biofilm on the membrane surface. After prolonged use, cleaning the membrane will not improve its performance. At that point the membrane has to be replaced. Surface attrition or erosion of the membrane also takes place, which is an irreversible process. Maximum operating temperatures for various membranes are given in Table 5.1 and operating them beyond that temperature could damage them.

High-pressure drop can cause the membrane material to fail. Most membranes, except CA membranes are very resistant to extreme pH values.

TABLE 5.1   Maximum operating temperatures of few membranes

Membrane	°C
CA	35
PVDF	95
PSO Composite	80
UF of carrageenan	80–90

Polyester is used as a backing in many membranes, and is a major bottle-neck since it limits the upper pH limit to 11.5. Most membranes are stable at low pH. Highly viscous liquids lead to high-pressure drop and the flux may decrease. After a filtration cycle, the plant should be flushed with a volume of water about three times larger than the internal volume of the system or, 5 L of water per m^2 of membrane area. During flushing, the concentrate is drained. Membranes reject dissolved solids but they work very poorly if the feed contains considerable amount of suspended solids, or if solids precipitate during the process. So prior to membrane filtration the fluid has to be pretreated to remove suspended solids, oxidizers and precipitates. Inhibiting or arresting microbial growth is very important if food products are handled in the membrane. Prefilter is placed in the upstream to prevent plugging or damaging of membranes by hard or sharp particles from the feed. So a typical membrane filtration system contains several pretreatment units.

5.1.1 MICROFILTRATION AND ULTRAFILTRATION

MF and UF are based on physical separation of the solid from the liquid. The pore size determines the efficiency to which dissolved solids and micro-organisms are removed. Substances that are larger than the pores in the membrane are completely removed and those that are smaller than the pores of the membrane are partially removed. MF removes all bacteria, and part of viral contamination. This is because viruses can attach themselves to bacterial biofilm. MF is used in sterilization of beverages and pharmaceuticals; clearing of fruit juices, wines and beer; biological wastewater treatment; effluent treatment; separation of oil/water emulsions; pre-treatment of water for nano filtration or RO and solid-liquid separation for pharma or food industries since UF completely removes viruses. These are used in dairy industry (milk, cheese); food industry (proteins); metal industry (oil/water emulsions separation, paint treatment) and textile industry.

5.1.2 NANO FILTRATION AND REVERSE OSMOSIS

NF is mainly used in drinking water purification process mainly for water softening, decoloring, micropollutant removal; remove coloring agents;

and multivalent ions. Removal of pesticides, nitrates and heavy metals from groundwater can also be performed with NF.

RO is based on the concept of osmosis. Two liquids containing different concentrations of dissolved solids when made to come in contact with each other will mix until the concentration becomes uniform. When these two liquids are separated by a semi permeable membrane, pure liquid containing a lower concentration of solid will move through the membrane into the liquid containing a higher concentration of dissolved solids. It is because the former has higher vapor pressure than the latter, and hence the movement is from higher pressure to lower pressure (Figure 5.3). After a while the water level will be higher on one side of the membrane. The difference in height is called the osmotic pressure.

In RO, the liquid is forced back through the membrane by applying an external pressure, while the dissolved solids remain in the same compartment. Osmotic pressure for seawater and pure water is 24 bar (350 psi). So by applying a pressure greater than 24 bar in the former, pure water can be pushed through the RO membrane but retaining only the salt. Hence RO is used in desalination. It is also used for preparing boiler feed water, concentrate fruit juice, sugar and coffee, concentrate wastewater and concentrate milk for cheese production. RO membranes can remove 95% of dissolved salts and 99% of bacteria. The pressure that needs to be applied is 2–17 bar for fresh and brackish water, and 40–70 bar for seawater.

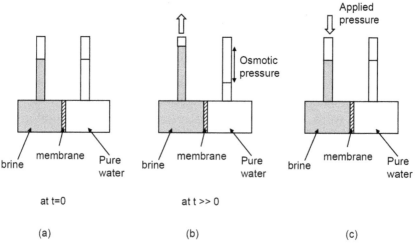

FIGURE 5.3   Concept of (a) and (b) osmotic pressure and (c) reverse osmosis.

The organic matter and the amount of bacteria should be minimal to prevent biofouling of membranes. Acids (such as, HCl or H_2SO_4), or anti-scalent is added to prevent scaling and precipitation of insoluble solids, such as, calcium carbonate and barium sulfate. Flotation or Filtration can be used to remove suspended solids.

5.1.3 COST

A Hybrid process consisting of chemical treatment and MF is cheaper than NF. But NF produces waters with a good quality. The effectiveness of chemical pretreatment depends on the organics present in the water. Cost of chemical pretreatment is greater than the energy costs for NF if the concentration of organics to be removed is high. If only removal of turbidity and micro-organisms are desired then MF is cheaper. Ferric chloride is generally used as a coagulant to remove suspended solids. Pretreatment cost becomes very high for MF when it is operated at low flux and high ferric chloride dosage, since large membrane area is required to overcome the low flux. Disposal of ferric chloride waste is also a major problem.

5.1.4 DESIGN EQUATIONS

Hermans and Bredee (1936) showed that the decrease in flux during constant pressure filtration is represented by the following equation

$$\frac{d^2t}{dV^2} = K\left(\frac{dt}{dV}\right)^n \tag{5.1}$$

where t is the filtration time, V the cumulative filtrate volume, K is a proportionality constant and its value depends on the filtration process.

For cake filtration model, in which fouling occurs due to the formation of a bed of particles on the surface of the membrane, then $n = 0$.

In the intermediate blocking model, where particles settle on existing foulant deposits, then $n = 1$.

For the pore constriction model, where the foulants are deposited on the inner surface of the pore leading to a decrease in the diameter of the pores, then $n = 1.5$, and if fouling occurs by complete blocking of the pore then $n = 2$.

dt/dV is the inverse of filtrate flow rate while the second derivative, d^2t/dV^2, is proportional to the rate of increase in the total resistance to filtration.

An analytical solution is developed by assuming a spatially uniform cake resistance (Chandler and Zydney, 2006) as follows

$$Q = Q_o \left[\exp\left(-\frac{\alpha \Delta P C_b}{\mu R_m} \right) + \frac{R_m}{R_m + R_p} \left(1 - \exp\left(-\frac{\alpha \Delta P C_b}{\mu R_m} \right) \right) \right] \qquad (5.2)$$

$$R_p = (R_m + R_{po}) \sqrt{1 + \frac{2 f \alpha_s \Delta P C_b}{\mu (R_m + R_{po})^2} t} - R_m \qquad (5.3)$$

where Q = volumetric flow rate, Q_0 = initial flow rate through the unfouled membrane, α = area blocked per mass of deposit, ΔP = transmembrane pressure drop, μ = fluid viscosity, C_b = bulk mass concentration of foulant, R_m = clean membrane hydraulic resistance, R_{po} = resistance of the initial foulant deposit, f = fraction of the bulk concentration that contributes to deposit growth, R_p = time-dependent resistance of the growing cake, and α_s = is the specific resistance of the foulant cake.

The osmotic pressure of the solution will play a role in the filtration process, and it will decrease the driving force (ΔP). So the flux will be

$$J = L(\Delta P - sP_{os}) \qquad (5.4)$$

where P_{os} is the osmotic pressure, and if the solute is completely rejected by the membrane then s=1, and depending upon the efficiency of the filtration process may vary between 0 to 1. As can be seen from the equation the second term decreases the driving force. For dilute solutions, osmotic pressure can be simplified to

$$P_{os} = RTc_s \qquad (5.5)$$

where c_s is the concentration of the solute that is being filtered near the surface of the membrane. This concentration will not be the same as what is present in the bulk of the solution, unless the upstream is thoroughly mixed. If R_m is the membrane resistance then $1/R_m$ is called the permeability (L). If s=0 then

$$J = \Delta P / R_m \qquad (5.6)$$

An equation for the flux of the solvent can be made as given below, assuming dilute concentration and all the solute is retained in the upstream.

$$\frac{dV}{dt} = -AJ = -AL\Delta P\left(1 - \frac{RTc_s}{\Delta P}\right) \qquad (5.7)$$

If we assume $C_s = n/V$, where n is the moles of solute and V = is the volume of the solvent retained. Then the equation can be integrated assuming at t=0, V=V_0

$$t = \frac{1}{AL\Delta P}\left(V_o - V + \frac{RTn}{\Delta P}\ln\left[\frac{V_o - RTn/\Delta P}{V - RTn/\Delta P}\right]\right) \qquad (5.8)$$

Average permeability (L m^{-2}h^{-1}bar^{-1}) for Polyamide-urea membranes = 3.8 and for Polyamide on polysulfone support = 11–15. Water permeability (L) for γ-alumina, $CoAl_2O_4$ and $ZnAl_2O_4$ are 1.2 5.7 and 3.5 respectively. Membranes are also made of Alumina, titania and zirconia (ATZ) (Manufacturer Tami Industries). The permeability of such membrane is (mL/(cm^2 min psi)) 0.0129, 0.0131 and 0.0055 for MW cut off of 50, 15 and 1 kDa respectively.

5.1.5 CONCENTRATION POLARIZATION

The concentration at the surface of the membrane will be much more than what is in the bulk, because as the solvent flows through the membrane there will be an accumulation of the solute at the surface in the upstream. This phenomenon-is called concentration polarization.

The solute carried towards the membrane surface (as flux = J) will be equal to material through the membrane

$$J = v\,C = -D\frac{dc}{dx}$$ (5.9)

where x is the direction along the flow and D is the diffusion coefficient. v is the velocity. The boundary conditions will be at $x = 0$, $C = C_s$ (near the wall on the upstream side) and at $x = d$, $C = C_b$; $d =$ thickness of the boundary formed and C_b is the concentration of the solute in the bulk (see Figure 5.4). Integrating this equation one arrives at a relationship between bulk and surface concentrations of the solute.

$$J = \frac{D}{d}\ln\frac{C_s}{C_b}$$ (5.10)

where D/d is a mass transfer coefficient relating this transport process. Pressure drop in the cross flow membrane filtration system is

$$\Delta P = \Delta P_m + \Delta P_p + \Delta P_c$$ (5.11)

where ΔP_m = pressure drop across the membrane, ΔP_p = across polarization layer, and ΔP_c = across cake layer.

5.2   DIALYSIS

Dialysis is suited for the separation of small solutes or salts from a mixture containing large biological macromolecules. The separation is due

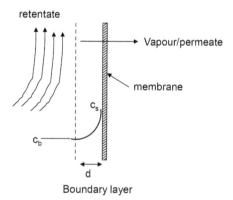

FIGURE 5.4   Accumulation of solute near the upstream surface of the membrane (concentration polarisation).

to the concentration gradient. Dialysis finds application in the treatment of patients with renal failure. Blood from an artery in the arm (or leg) is drawn into a dialysis machine and the purified blood is fed back into a vein. The patient has to undergo this process once every two or three days and it may last 6–8 h. The two major chemicals in the urine that has to be in control are creatinine and blood urea nitrogen and if kidneys are not functioning properly their levels go up in the blood.

There are two types of dialysis namely, hemodialysis and peritoneal dialysis. The former uses a membrane to remove excess waste products from the body. The latter uses a fluid that is placed into the patient's stomach cavity through a tube to remove the excess waste products from the body. This fluid is then removed from the body.

During hemodialysis, blood passes from the patient's body through a dialyzer (Figure 5.5). The dialyzer contains thousands of small membrane fibers. The Dialysis solution is pumped outside these fibers. The salts from the blood flow through the membrane and reach this solution, which carries away the waste.

The dialyzer clearance (K_d, mL/min) is the removal rate of a substance from the blood and is given as

$$K_d C_i = Q_i C_i - Q_o C_o \qquad (5.12)$$

where Q_i and Q_o are the blood flow at the inlet and outlet of the dialyzer, respectively. C_i and C_o are the concentrations of the substance that is removed in and out (mmol/L). $C_o \ll C_i$

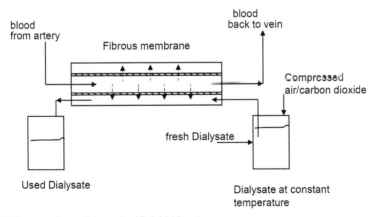

FIGURE 5.5   Hemodialyser (artificial kidney)

If Q_F is the ultrafiltration rate (i.e., some blood fluid also escapes through the membrane) then,

$$Q_i = Q_o + Q_F \tag{5.13}$$

Then K_d becomes,

$$K_d = Q_i - (Q_i - Q_F)C_o / C_i \tag{5.14}$$

In hemodialysis, $Q_i = Q_o = Q_B$ and $Q_F = 0$, (i.e., no blood fluids escapes) then

$$K_d = Q_B(1 - C_o / C_i) \tag{5.15}$$

5.3   ELECTRODIALYSIS

Here separation of salts/electrolytes is achieved by applying a voltage. The unit consists of an anode and a cathode at two ends with a series of membranes and separators placed parallel to one another, like in a plate and frame filter press. In a commercial plant about 50–250 such units may be present. They have a high electrical resistance leading to high power consumption. The electrical voltage applied across the electrodes drives the ions through the membranes. The membranes can selectively reject ions of certain charge and allow ions of opposite charge (Figure 5.6). Anions will pass through anion

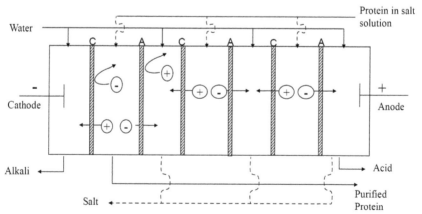

FIGURE 5.6   Electrodialysis system (C-Cation permeable membrane, A – Anion permeable membrane).

membranes and cations pass through cations membranes. They can be used to concentrate a desired ion or reject unwanted ions, namely to remove electrolytes from protein solution, desalting of seawater, food processing, recovery of metals, and wastewater treatment in electroplating industries. If an ionic product in the fermentation medium is inhibiting then it can be removed continuously using an electrodialysis system and the broth can be recycled back.

Several types of electrodialysis cells are present:

1. desalting cell;
2. two compartment cation;
3. two compartment anion; and
4. three- compartment bipolar.

There are two types of membranes, (i) ion permeable, and (ii) bipolar. The former membrane consists of either cation or anion exchange resins. A cation-exchange membrane will have negatively charged groups such as, SO_3^- covalently bound to a polymer backbone chain such as, styrene and divinylbenzene copolymers. Ions with a charge opposite to the fixed charge namely counter ions freely pass. The concentration of counter ions such as, Na^+ will be relatively high; so they carry the electric current through the membrane. The negative charges attached to the polymer chains repel ions of the same charge (co-ions), namely anions and so their concentration through the membrane will be relatively low. Attachment of positively fixed charges (e.g., $-NR_3^+$ or $C_5H_5N^+R$ where $R = CH_3$) to the polymer chain leads to anion permeable membranes. This will selectively transport negative ions, and will repel positive ions (cations). This exclusion, as a result of electrostatic repulsion, is called 'Donnan exclusion'.

Polymers such as, polystyrene sulfonic acid are water soluble, so they are cross-linked with compounds such as, divinylbenzene. The properties of the membrane are adjusted through (i) degree of cross-linking, (ii) charge density, and (iii) mechanical stability. Higher level of crosslinking improves selectivity and membrane stability by reducing swelling, but it also increases electrical resistance. High charge density reduces resistance and increases selectivity. It leads to increase in swelling and hence requires more crosslinking. A compromise between selectivity, electrical resistance, and mechanical stability has to be arrived at by manipulating the crosslinking density and charge density of the polymer.

Desalting cell comprises of a cation membrane, feed vessel, an anion membrane and a concentrate or product vessel. This cell is located near the anode end. This unit cell finds application in the desalination of dilute acetic acid, and in the recovery and production of salts of ammonia.

Two-compartment cation cell consists of a bipolar membrane, a feed or salt/acid tank and a cation membrane. For example, if ammonium acetate is fed into the compartment, it is acidified by the bipolar membrane. A cation membrane transports the ammonium cation. In a base or product compartment this ammonium cation combines with the OH ions generated by the bipolar membrane to form ammonium hydroxide.

Two-compartment anion cell consists of a bipolar membrane, a product or an acid compartment, an anion membrane and a feed or salt/base compartment. This cell is used to basify the salt. For example an ammonium salt solution can be basified to generate an acid product and an ammonia rich base solution. Three-compartment bipolar cell comprises of a feed or salt compartment, a cation membrane, a base compartment, a bipolar membrane, an acid compartment and an anion membrane. For example ammonium acetate salt can be converted into ammonium hydroxide and acetic acid. Cells containing more than three membranes and three compartments are also available.

Bipolar membranes (as the name implies) consist of an anion and a cation permeable membrane joined together. When this bipolar membrane is oriented such that the cation-exchange layer faces the anode, on application of a potential field across the membrane, splitting of water into proton and hydroxyl ions will occur. This generates acidic and basic solutions at its surface. There are several advantages to water splitting with bipolar membranes and they include (i) gasses are not evolved at the surface or within the bipolar membranes, (ii) energy in the conversion of water to O_2 and H_2 is saved, (iii) power consumption is about half that of electrolytic cells, and (iv) The electrodes used in conventional electrolytic cells, are expensive while these membranes are inexpensive.

Acid and base can be produced from a neutral salt by placing multiple bipolar membranes along with ion permeable membranes in between a single pair of electrodes in an electrodialysis stack. This approach is inexpensive and generate minimum amount of unwanted byproducts.

The governing mathematical relation in electrodialysis is (LaGredaetal, 1994)

$$I = \frac{FNQ\eta}{nE} \tag{5.16}$$

where F = Faraday's constant = 96487 coulombs/g-equivalent; I = current; N = solution normality; Q = flow rate, m^3/s; η = removal efficiency; n = number of cells; E = current efficiency.

Power required for electrodialysis is given as

$$P = I^2 R \tag{5.17}$$

where R is the resistance.

5.4   PERVAPORATION

Pervaporation is a combination of membrane permeation and evaporation. It has several advantages over the normal evaporation and they include (i) low temperatures and pressures, (ii) lower cost and better performance than any method for the separation of azeotropes, (iii) economical for the dehydration of organic solvents, (iv) ideal for removal of organics from aqueous streams, (v) no entrainer is required, (entrainer is required for breaking an azeotropic mixtures) hence no contamination of the product, (vi) the operation is independent of vapor – liquid equilibrium(distillation depends on vapor – liquid equilibrium), (vii) can be performed in batch or continuous mode, and (viii) suited for separation of heat sensitive products (in food or pharmaceutical products).

In this process, the separation of two or more components is achieved with the help of a thin polymer membrane, which has different rates of diffusion. Evaporation of the mixture is achieved by heating. A vacuum is applied on the permeate side and the vapor is condensed. Figure 5.7 shows an overview of the pervaporation process. The permeate must be volatile at the operating pressure and temperature and the solute must be compatible with the polymer.

Pervaporation can be performed in batch or continuous mode. The former is a simple system and is flexible. A tank is required to hold the product in the batch operation. Continuous pervaporation consumes very little energy and product can be obtained continuously. For mixtures with low

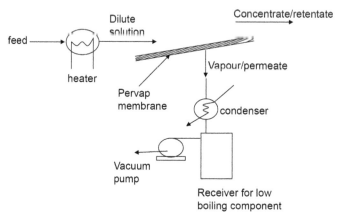

FIGURE 5.7   Pervaporation separation system.

impurities and large volume, this mode of operation is very good and less energy intensive than distillation.

The steps in pervaporation process include (i) evaporation of the permeate, (ii) sorption of the permeate at the interface of the membrane, (iii) diffusion across the membrane due to the concentration gradient (rate determining step), (iv) desorption into the vapor phase on the permeate side of the membrane, and (v) condensation of the permeate as liquid. The second and third steps are responsible for the permselectivity. As material passes through the membrane it swells, making it more permeable, but less selective. When the selectivity drops down to a very low value, the membrane must be regenerated.

The other driving force for separation is the difference in partial pressures across the membrane. This can be achieved by two ways namely by (i) using vacuum or (ii) sweep gas pervaporation. In the former method reducing the pressure on the permeate side (by applying vacuum) of the membrane, creates a driving force. In the second method a partial pressure gradient is induced by sweeping an inert gas over the permeate side of the membrane across so that it will carry the vapor. The liquid product collected is rich in the more readily permeating component of the feed mixture. The retentate is made up of the feed that cannot pass through the membrane.

The membranes used in pervaporation processes are divided as hydrophilic and organophilic (hydrophobic). The former membranes are used to remove water from organic solutions. They are made up of polymers with glass transition temperatures above room temperatures. Polyvinyl alcohol allows water to pass through. Hydrophobic membranes are used to recover organics from solutions. These membranes are made up of elastomers (polymers with glass transition temperatures below room temperature) that are flexible. Examples include nitrile, butadiene rubber, and styrene butadiene rubber. For the removal of trichloroethylene from water the membranes are made up of dense silicone rubber. NaY zeolite membrane is used for the separation of methanol–methyl tert-butyl ether mixtures. Membranes made from inorganic materials are superior than organopolymeric materials with respect to thermal and mechanical stability and chemical resistance. The latter also does not swell unlike the former. Zeolite crystals have several properties including molecular sieving, ion exchange and selective filling of water in the micropore.

Liquid movement in pervaporation is described by solution-diffusion model. Higher flux can be obtained with an increased thermal motion of the polymer chains and the diffusing species. Properties of the polymers that affect diffusion include (i) backbone material, (ii) degree of cross-linking, and (iii) porosity. Molecular-level interactions between membrane and diffusing species are expressed in the form of an Arrhenius relationship:

$$P = P_o e^{-E/RT} \tag{5.18}$$

where, E = activation energy; P_o = permeability constant; R = gas constant; T = temperature in K.

Molecular flux (J_i), which is the amount of a component permeating through a membrane per unit area per unit time is given as

$$J_i = Q_i / At \tag{5.19}$$

where J_i = flux of component i (moles/h cm²); Q_i = moles of component i permeated in time t; A = membrane surface area (cm²).

Permselectivity is a parameter that describes the performance of a given membrane and it is defined as,

$$\alpha = \frac{(X_i^{perm} / X_j^{perm})}{(X_i^{feed} / X_j^{feed})} \qquad (5.20)$$

Assuming the density of the components in the feed is the same, then:

$$\alpha = \frac{(V_i^{perm} / V_j^{perm})}{(V_i^{feed} / V_j^{feed})} \qquad (5.21)$$

where X = weight fraction of each component; V = volume fraction of each component; ρ = density.

Superscripts 'perm' and 'feed' denote permeate and feed respectively while i and j represent the two components present in the feed.

The molecular flux across a membrane can be related to the permeability coefficient by:

$$J_i = -P_{L,i}(P_i^o X_{f,i} - \Delta P Y_{p,i}) / b_m \qquad (5.22)$$

where $P_{L,i}$ = permeability coefficient of component i; ΔP = change in partial pressure of pure component i across the membrane; $X_{f,i}$ = mole fraction of component i in the feed liquid; $Y_{p,i}$ = mole fraction of component i in the permeate; b_m = membrane thickness; P_i^o = pressure.

Industrial applications of pervaporation include (i) treatment of wastewater contaminated with organics, (ii) recovery of valuable organic compounds from process streams, (iii) separation of 99.5% pure ethanol-water solutions, (iv) breaking azeotropes such as, ethanol-water, isopropanol-water, THF-water, (v) continuous water removal from condensation reactions such as, esterification's to enhance conversion (shifting the equilibrium to the right), and (vi) combining distillation and vapor permeation.

Pervaporation can be combined with distillation to break azeotropes and prepare very dry solvents. Figure 5.8 shows two different distillation cum pervaporation assemblies for separating water and a solvent, in the first case the solvent is lighter than water and in the second case the solvent is heavier than water. In both the cases they form an azeotrope. The general approach that is traditionally practiced to break azeotropes includes (i) add a third solvent (entrainer) to break the azeotrope. This leads to contamination of the product and also adds to the operating cost, (ii) change of pressure, and (iii) use of adsorbent to selectively remove

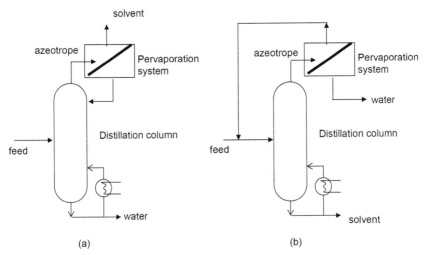

FIGURE 5.8 Pervaporation cum distillation system for separating water and solvent (a) for light solvent (b) for heavy solvent.

one liquid. Pervaporation could be a very cheap and less energy intensive alternative.

5.5 ISOELECTRIC FOCUSING

Isoelectric point (IP) is the value of pH at which the charge on the molecule is zero. Isoelectric focusing (IEF) represents the first dimension in the two-dimensional (2D) electrophoresis. It takes place in a pH gradient and it is suitable for separating molecules, which can acquire positive or negative charge including proteins, enzymes and peptides. The pH gradient is produced by an electric field. Before an electric field is applied, the gel has an uniform pH-value and almost all the compounds are charged (Figure 5.9). When an electric field is applied, the negatively charged compounds move towards the anode, and the positively charged ones to the cathode. Their velocity depends directly on the magnitude of their net charge and indirectly on the size. As soon as they reach their IP, they stop moving. Isoelectric focusing does not denature the protein, and hence is a good separation technique. The barrier used in IEF is a bipolar membrane or a combination of anion and cation selective membranes. The IEF setup contains several reservoirs. Adjacent reservoirs are separated by selective

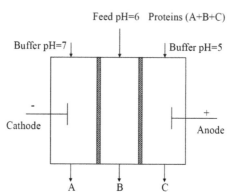

FIGURE 5.9   Isoelectric focusing setup (Iso pH for A=7, Iso pH for B=6, Iso pH for C=5).

permeable membrane, which allow interchange of solutes/proteins, but inhibit bulk fluid flow. Problems in this technique are (i) the continual remixing of purified materials with semi-purified or crude starting materials, and (ii) the heat that is generated during the process which needs to be dissipated to prevent heat buildup.

The velocity of the charged particle in an electric field is directly proportional to the voltage (ΔV) and the charge on the protein (C_I) and inversely proportional to the distance between the electrodes (d_e) and size of the protein. So

$$v = (D_e/RT)C_I\ \Phi\ \Delta V/d_e \tag{5.23}.$$

where D_e is the diffusion coefficient of the protein and is inversely proportional to the size of the protein and viscosity of the solution; and Φ is a constant = 96,500 current/mol.

Isoelectric pH values of some common proteins are listed in the Table 5.2.

Dilute acids or bases are used as the electrolyte solutions in the electrode chambers, 0.1 M of sodium hydroxide (pH 12.5) for the catholyte and 0.1 M of phosphoric acid (pH 2.3) for the anolyte is generally preferred. As the IEF takes place, the pH gradient will be between these two limits. The extreme ends adjacent to the electrolyte solutions buffer to the electrolyte pH. This phenomenon-is known as 'spiking.' When the number of chambers is small, spiking would reduce the resolution.

TABLE 5.2 Molecular weight and isoelectric pH values of few proteins

	Molecular weight	Isoelectric pH
Gelatin	10,000–100,000	4.8–4.85
Insulin	40,900	5.3–5.35
Cytochrome C	15,600	9.7
Myoglobin	17,200	7.0
Urease	480,000	5.0–5.1
Hemoglobin	66,700	6.79–6.83

5.6 LIQUID–LIQUID EXTRACTION

The principle of extraction is based on the partioning of the solute between two immiscible liquids, such as, water and a hydrophobic solvent. When separation by distillation is ineffective, very difficult, or temperature sensitive materials need to be separated, then liquid–liquid extraction is one of the best alternatives to be considered. Close-boiling mixtures that cannot withstand high temperature may be separated by extraction. Penicillin is recovered from fermentation broth by extraction with butyl acetate. Citric acid is more soluble in methyl amyl ketone than water at a pH of 4. So it can be used to extract the acid from an aqueous solution. This technique can be used to selectively remove impurities or concentrate a particular chemical.

A solute (S) which is in equilibrium with the light (L) and heavy phases (H) can be represented as

$$S_H \leftrightarrow S_L$$

$$\Delta G = \Delta G^\circ + RT \ln (a_L) - RT \ln (a_H) \tag{5.24}$$

where a = activity and G = free energy.

$$\Delta G = \Delta G^\circ + RT \ln [(a_L)/(a_H)] \tag{5.25}$$

The partition coefficient, $Kp = (a_L)/(a_H) = [S_L]/[S_H]$

where $[S_H]$ = concentration of solute in the heavy; and $[S_L]$ = concentration of solute in the light. For example, the heavy is the aqueous fermentation broth and the solvent is the light phase.

We can use concentrations if the solutions are dilute and follow ideal behavior.

At equilibrium $\Delta G=0$, then

$$K = e^{-\Delta Go/RT} \tag{5.26}$$

If the solute is in different forms, such as, ionized, non-ionized, etc., then another term called distribution coefficient defines this partition and it is defined as

$$Dc = \frac{\text{(total of all equilibrium forms of the solute in the solvent phase)}}{\text{(total of all equilibrium forms of the solute in the heavy phase)}}$$

This ratio depends on pH. Assume a solute in the aqueous phase (H) ionizes as

$$HA \leftrightarrow H^+ + A^-$$

and the ionization constant,

$$K_a = [H^+] [A^-]/[HA]. \tag{5.27}$$

In the solvent phase (phase L) the solute does not ionize then,

$$K_p = [HA]_L/[HA]_H \tag{5.28}$$

$$K_a = [H^+]_H [A^-]_H/[HA]_H \tag{5.29}$$

$$Dc = [HA]_L/\{[HA]_H + [A^-]_H\} = K_p [H^+]_H/\{[H^+]_H + K_a\} \tag{5.30}$$

At highly acidic conditions $[H+]_H \gg K_a$, then $Dc = K_p$, that is, It is independent with respect to pH (zero order dependence).

At alkaline conditions $[H^+]_H \ll$, then $Dc \, \alpha \, [H^+]_H$, that is, Dc is directly proportional to pH (first order dependence).

Extraction systems deviate from ideal behavior due to (i) dissolution of one phase into another, (ii) solute saturating in a phase, (iii) reaction of solute with the solvent, and (iv) alteration of pH and other operating conditions during the extracting process.

As per Gibb's phase rule,

$$P+d=N+2 \qquad (5.31)$$

where P = number of phases, N = number of components, and d = degrees of freedom. In the liquid–liquid extraction process if we have two phases and one component (solute), then d = 1 (degrees of freedom is only one).

Neutral organic solutes can be extracted from water, base or acid using an appropriate organic solvent. Several points that need to be kept in mind while selecting the solvent are (i) free base is more soluble in non-polar organic solvents than in polar organic solvents, water or aqueous base; (ii) quaternary ammonium salts of amines (formed by reaction with acids) are more soluble in polar or aqueous media than non-polar organic solvents; (iii) if the aqueous solvent is basic, amine compounds may be in the free base form; (iv) free acid may be more soluble in an organic solvent than in water or acid; (v) salts formed by the reaction with base are more soluble in aqueous media than in organic solvents; (vi) acidic aqueous medium will be in the free acid form and so will partition into organic phase; and (vii) n-butanol, ethyl acelate, MIBK, Toluene and Hexane are a few of the solvents that can be used to extract acetic acid from water. n-Butanol has the highest and n-hexane has the lowest extraction ability while n-butanol has the highest and n-hexane has the lowest miscibility with water.

Optimization of the operation of the extraction processes involves:

(a) Solvent selection, which is the most important challenge in L-L extraction. The solvent should (i) extract the desired component from the mixture and not the other components. (ii) should have high distribution coefficient and good selectivity towards solute, (iii) little or no miscibility with feed solution (iv) easily recoverable (v) non-toxic (vi) cheap (vii) physico-chemical properties such as, boiling point, density, interfacial tension, viscosity, corrosiveness, flammability, stability, compatibility with product, availability should be acceptable.

(b) Appropriate operating conditions, which includes temperature, pressure, ratio of solvent to heavy, pH, residence time. High temperature reduces the viscosity and decreases the mass-transfer resistance. In extraction of products including pencillin or agrochemicals such as, orthene, pH is maintained constant to increase distribution coefficient and minimize product degradation. Ethyl acetate may hydrolyze in the presence of mineral acids to acetic acid and ethanol. Long residence time affects short-life components (e.g., antibiotics and vitamins).

(c) Mode of operation, namely batch, continuous or semi-continuous; co-counter or cross current; mode of operation determines the solvent requirement and size of the vessels.

(d) Extractor type – several designs are reported in literature each one has its own positive and negative features.

(e) Hardware design determines the capital and operating costs.

A single-stage extractor can be represented as shown in Figure 5.10. The solute mass balance is given as,

$$F\,X_f + S\,Y_0 = F\,X_r + S\,Y_1 \qquad\qquad (5.32)$$

where F = feed quantity; X_f, X_r, Y_0, and Y_1 are the weight fractions of solute in the feed, raffinate/solvent and extract, respectively.

The partition coefficient $K = Y_1/X_r$ at equilibrium   (5.33)

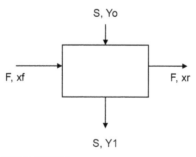

FIGURE 5.10   Single stage extractor.

The assumption here is immiscibility of feed and solvent. If initial solvent is free of solute, (i.e., $Y_0 = 0$) then,

$$X_r = X_f (F/(F+SK)) \qquad (5.34)$$

$$S = (F/K) ((X_f/X_r)-1) \qquad (5.35)$$

Equation (5.34) indicates the decrease in solute concentration after the extraction process. Equation (5.35) indicates the amount of solvent required to decrease the solute concentration from X_f to X_r.

Batch extractors are good for low capacity and multi-product plants including pharmaceutical, specialty and agrochemical industries. For a few stage operation crosscurrent is practical, and economical and flexible, but crosscurrent with multiple solvent inlet points (Figure 5.11) leads to high solvent usage. Batch extractor is usually an agitated tank containing two vessels, one an agitated vessel for mixing and another for settling (having a large diameter). The residence time in the second vessel will be much higher than in the first vessel. Continuous extractors are more compact than other designs.

For multi-stage crosscurrent operation, where equal amount of solvent is added at each stage (Figure 5.11), assuming that the partition coefficient (K) is constant over the concentration range and fresh solvent does not contain any solute then

$$X_r = X_f (1/(1+E_F))^n \qquad (5.36)$$

$$S = F/K [(X_f/X_r)^{1/n}-1] \qquad (5.37)$$

KS/F is defined as the extraction factor (E_F). For larger volume of extraction and for efficient use of solvent, countercurrent mixer-settlers are employed (Figure 5.12). Countercurrent operation conserves the mass transfer driving force.

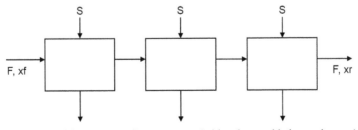

FIGURE 5.11   Multi stage cross flow extractor (with solvent added at each stage).

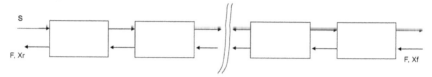

FIGURE 5.12   Multi stage countercurrent extractor.

It can be shown that for a 'n' stage countercurrent operation, where the solvent is flowing in one direction and the mixture is flowing in another direction (see Figure 5.12), the raffinate concentration at the end of n stages would be

$$X_r = X_f \left[\frac{(E_F - 1)}{(E_F^{n+1} - 1)} \right] \tag{5.38}$$

One could estimate the number of stages required to achieve a reduction in the solute concentration from X_f to X_r by rearranging Eq. (5.38) to get

$$n = \frac{\ln\left[(E_F - 1)\dfrac{X_f}{X_r} + 1 \right]}{E_F} - 1 \tag{5.39}$$

Higher is the E_F, higher is the reduction ratio. Systems with E_F less than 1.3 will not be commercially viable. For a given extraction factor (E_F), and number of stages (n), the countercurrent mode of operation out performs the crosscurrent mode (see Figure 5.11).

If the feed is introduced in the k-th stage as shown in the Figure 5.13, but still the liquids flow in countercurrent fashion then we need to write two equations for the stages 1 to (k-1) and for stages (k+1) to n independently. These two equations will be connected by the mass balance in the k-th stage. Then,

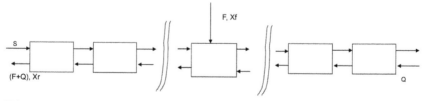

FIGURE 5.13   Multi stage cross flow extractor (with solvent added in only one stage).

$$X_F = E_F X_r \left[\frac{((E_F')^k - 1)}{((1/E_F)^{n-k+1} - 1)} \right] \left[\frac{(1/E_F - 1)}{(E_F' - 1)} \right] + (1 + Q/F) X_r \quad (5.40)$$

These equations assume in addition to previous assumptions that (i) K is constant in all the stages, (ii) equilibrium is maintained in each stage, and (iii) no change in physical properties of various fluids. The two extraction factors E_F and E_F' are defined as

$$E_F = KS/F \text{ and } E_F' = KL/(Q+F) \quad (5.41)$$

Commercial extractors can be classified into four broad categories as (i) mixer-settlers, (ii) centrifugal devices, (iii) static column contactors, and (iv) agitated column contactors. Each has its own advantages and limitations as condensed in the Table 5.3. Flow rate and residence time in mixer-settler is high when compared to other designs. They also occupy more floor space when compared to agitated or static columns.

Mixer-Settler, (Figure 5.14) as the name indicates, consists of a series of agitated vessel followed by a settling tank. The intimate mixing takes place in the former and the separation is achieved in the latter. Sufficient residence time has to be given in the latter vessel. This operation is repeated with fresh solvent washes or with partially used solvent. This unit occupies large floor space.

Column extractors (Figure 5.15) include static column, spray column, sieve plate column, random or structured pack column, agitated column, rotating disc contractor, reciprocating agitated column and pulsed column. Scheibel, Kiihri and Karr columns (Figures 5.16 and 5.17) are a few specialized column extractors. Countercurrent

TABLE 5.3 Comparison of various extractors

	Stages it has	Flow rate	Residence physical properties of		Floor area occupied
			time	fluid it can handle	
Mixer settler	L	H	H	L-H	H
Centrifugal	L	L	L	L-M	M
Static column	M	M	M	L-M	L
Agitated column	H	M	M	L-H	L

(L-Low, M-Medium and H-High).

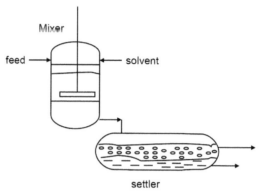

FIGURE 5.14   Mixer settler batch extractor.

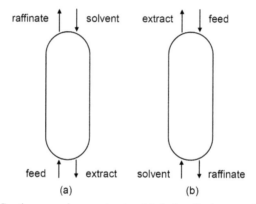

FIGURE 5.15   Continuous column extractor (a) design for heavy solvent (b) for light solvent.

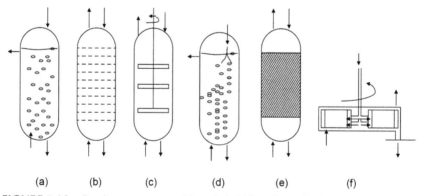

FIGURE 5.16   Continuous extractor (a) static bubble column (b) sieve plate column (c) agitated column (d) spray column (e) packed column (f) rotating disc

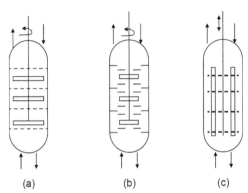

FIGURE 5.17   Continuous extractor (a) Kuhni column (b) Scheibel column (c) Karr column.

column contactors are popular rather than concurrent. Packing or internals inside the extractor brings efficient contact, create turbulence, and produce finer droplets. Agitated columns bring higher degree of mixing, but lead to excess energy usage. Axial mixing (along column length) in column extractor reduces stage efficiency. Baffles are used to minimize axial mixing in columns. Poor mixing leads to the formation of large droplets and decreases interfacial area for mass transfer. Temperature gradients in columns have to be avoided. Columns occupy the least floor space.

Centrifugal extractors are high-speed rotary units with very low residence time and very high efficiency (90%). Mixing and separation times range from 10 to 30 seconds each. The number of stages in a centrifugal device is one, but currently devices with multiple numbers of stages are also in the market. These extractors are mainly used in pharmaceutical industry.

Factors which need to be considered when selecting the type of extractor are (i) number of stages, (ii) fluid properties, (iii) operational considerations, (iv) presence of solids, (v) safety, (vi) ease of maintenance, and (vii) cost.

Emulsions are formed due to over agitation, due to the inherent nature of the chemical compounds or due to contaminants that lower the interfacial tension. Coagulants are added to prevent emulsification. Passing the emulsion through a coalescer can break them. A layer containing loose solid impurities that float at the interface is called a rag layer. This needs to be collected separately, filtered and recycled.

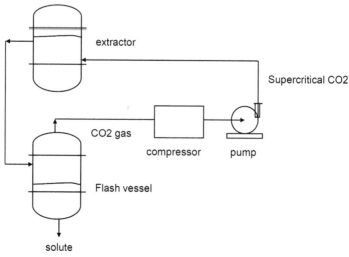

FIGURE 5.18   Supercritical extractor system.

5.7   SUPERCRITICAL EXTRACTION

Supercritical extraction is based on based on the use of supercritical fluid as an extracting medium. Such fluids have the advantages of a dense gas or a light liquid. Here, the recovery is easy, since by reducing the pressure or temperature the solute could be separated from the fluid as gas, which can be recycled again (Figure 5.18). It is a green approach and uses less energy during the recovery stage. CO_2 is a versatile supercritical fluid which behaves as supercritical fluid at 31°C and a pressure of 74 bar. This technique is used in decaffeination of coffee and tea, extraction of oils and fragrance from plants, spices and pepper. Supercritical water is also becoming popular (221 bar and 374°C), but the material of construction is a challenge since the pressure and temperature requirements here are very high.

5.8   SOLID–LIQUID EXTRACTION

Solid–liquid extraction or leaching is the process of removing solutes from a solid by using a solvent. Here, the solute partitions between the solid and the solvent phases. It is widely used in industries where mechanical or thermal methods of separations are not possible or if the solute is thermally liable. Extraction of sugar from sugar beets, oil from oil seeds, aromatic components from natural products/plants are a few industrial examples of leaching. This process is also known as lixiviation or washing. Removal of ash from coal prior to charging the latter into the boilers is called as washing.

During leaching the solvent diffuses through the pores of the solid. The solute dissolves (partitions) into the solvent and this solution diffuses back to the bulk. The solute is then recovered from this solution by distillation or evaporation and the solvent is reused or recycled. The solid particles are kept in a packet bed while the solvent flows through this bed. After sufficient contact time, the mixture is discharged and sent to a filter. The leaching process could be carried out in multiple stages similar to liquid-liquid extraction, where the solvent could be flowing with respect to the solids, in co or countercurrent manner.

The solvent should have low viscosity and good diffusion properties so that it can percolate through the pores of the solid matrix. It should not be toxic, inflammable and easily recoverable.

5.9 ADSORPTION

Adsorption is an equilibrium process and the calculations are based on mass balance and partition of solute between the fluid and the solid. Here, a solid phase is also present in addition to the liquid/gas phase. Adsorption mechanisms are categorized as physical adsorption or chemisorption. Weak molecular forces, such as, van der waals, ionic and electrostatic forces play a role in adsorption. Activated carbon, silica gel, zeolite and alumina are a few of the common adsorbents. Three common adsorption isotherms are linear, Freundlich and Langmuir (Figure 5.19).

The linear isotherm is given by

$$q = Kx \tag{5.42}$$

The Freundlich isotherm

$$q = Kx^n \tag{5.43}$$

The Langmuir isotherm is

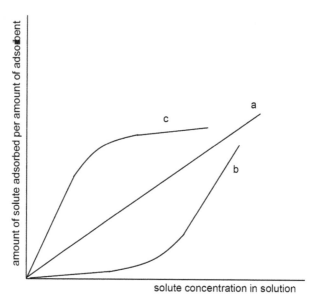

FIGURE 5.19   Various adsorption isotherms (a) Linear (b) Freundlich (c) Langmuir.

$$q = \frac{q_o x}{K_e + x}$$

(5.44)

where q = amount of solute adsorbed per amount of adsorbent (gm/gm); x = solute concentration in solution (gm/mL); K = equilibrium constant; n = constant (if the adsorption is favorable, then n < 1; if it is unfavorable, then n > 1); q_o and K_e = constants (q_o maximum adsorption capacity and Ke adsorption constant).

In the Langmuir isotherm, at low solute concentration (low x), the behavior will be like a liner isotherm (first order) and at high solute concentration (high x), a saturation behavior will be observed (zero order). Langmuir isotherm assumes that the adsorbent has a fixed capacity.

5.9.1   BATCH ADSORPTION

Here the solution containing the solute and the adsorbent are added into a vessel, mixed thoroughly and separated by filtration. The solute (adsorbate) will partition into the adsorbent and reach an equilibrium. This is a time consuming process. The mass balance for the solute is given as

$$x_F F + q_F W = x F + q W$$

(5.45)

where x and x_F are the concentration of the solute in the final and feed solution, respectively; q and q_F are the final and feed concentrations of the solute in the adsorbent, respectively; F is the amount of feed solution; and W is the weight of adsorbent added. If the adsorbent is fresh then q_F will be $= 0$.

Then, the mass balance relation is rearranged to obtain

$$q = q_F + \frac{F}{W}(x_F - x) \qquad (5.46)$$

This equation relates q and x. With the help of Eq. (5.46) and one of the equilibrium relationships (5.42)–(5.44) one could estimate the concentration of the solute in the two phases (solid and liquid) at equilibrium after the adsorption process (point of intersection in Figure 5.20a–5.20c).

5.9.2 CONTINUOUS STIRRED TANK ADSORPTION

In this process the adsorbent is suspended in a tank and the solution containing the solute enters the tank from one end and leaves from another end (Figure 5.21). Adsorption of the solute to the adsorbent takes place

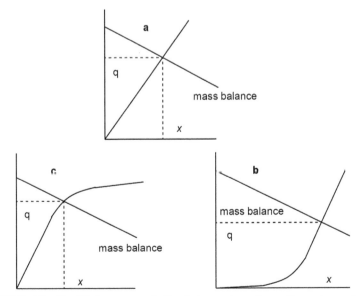

FIGURE 5. 20 Equilibrium concentration between two phases for various adsorption isotherms (a) Linear (b) Freundlich (c) Langmuir (a graphical approach).

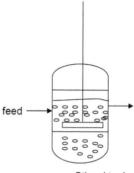

feed

Stirred tank

FIGURE 5.21   Continuous stirred tank adsorber.

during its stay in the tank. This approach can help in removing the solute in a continuous manner. Once the adsorbent is saturated it needs to
be removed and regenerated.

The mass balance for the solute in the liquid in a continuous stirred
tank absorber is

$$\varepsilon V \frac{dx}{dt} = F(x_F - x) - (1 - \varepsilon)Vr_{ads} \tag{5.47}$$

where V is the volume of the tank; x and x_F are the solute concentrations in
the outlet and the feed, respectively; F is the feed flow rate; r_{ads} is the rate
of adsorption per volume of tank; q is the adsorbed solute concentration;
and ε is the voidage. $(1-\varepsilon)$ is the volume fraction of the solids. A similar
mass balance on the adsorbent gives

$$(1 - \varepsilon)Vr_{ads} \frac{dq}{dt} = (1 - \varepsilon)Vr_{ads} \tag{5.48}$$

The controlling mechanism responsible for the kinetics of adsorption
can be:

1. diffusion from the solution to the adsorbent.
2. diffusion and reaction within the adsorbent particles (if the solute
    reacts after adsorption).

When diffusion in the solution controls the adsorption, then the rate r
is given by

$$r_{ads} = k_L a(x - x^*) \tag{5.49}$$

where k_L is the mass transfer coefficient, a is the surface area of adsorbent per tank volume, and x^* is a concentration in the solution which would be in equilibrium with the adsorbent.

The adsorption can follow one of the equilibrium relations as shown in Eqs. (5.42)–(5.44).

If we assume that the adsorption isotherm is linear then

$$q = Kx^* \qquad (5.50)$$

we then can integrate the Eqs. (5.47) and (5.48) assuming diffusion in liquid is controlling Eq. (5.49) to give the following analytical solution,

$$\frac{q}{Kx_F} = 1 - \frac{\sigma_1 e^{-\sigma_1 t}}{(\sigma_1 - \sigma_2)} + \frac{\sigma_2 e^{-\sigma_2 t}}{(\sigma_1 - \sigma_2)} \qquad (5.51)$$

$$\sigma_{1,2} = 0.5\left[b \pm \sqrt{b^2 - \frac{4k_L aF}{K(1-\varepsilon)V}} \right] \qquad (5.52)$$

and

$$b = \frac{F}{V\varepsilon} + k_L a \left(1 + \frac{\varepsilon}{K(1-\varepsilon)} \right) \qquad (5.53)$$

A similar equation for x can also be obtained.

5.9.3 FIXED BED ADSORPTION

Here the adsorbent is loaded in a tube (Figure 5.22). Fluid containing the solute flows into one end of the tube, and flows out from the other end. Initially, most

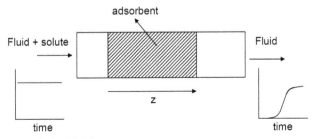

FIGURE 5.22 Tubular adsorber.

of the solute is adsorbed by the adsorbent, so that the solute concentration in the exit (effluent) is low. As adsorption continues (and adsorbent starts getting saturated with the solute), the effluent concentration in the exit rises, slowly at first, but then abruptly. This is known as breakthrough. Then the flow is stopped and the adsorbed material is eluted by washing the bed with the same solvent.

Adsorption in fixed bed is described by the mass balance on the solute in the liquid as:

$$\varepsilon \frac{dx}{dt} = -\upsilon \frac{dx}{dz} + D_d \frac{\partial^2 x}{\partial^2 z} - (1-\varepsilon)\frac{\partial q}{\partial t} \tag{5.54}$$

where ε is the void fraction in the bed, υ (=F/A) is the superficial velocity, A is a cross sectional area of the tube and D_d is a dispersion coefficient. The left hand side represents accumulation in the liquid. The first term on the right hand side corresponds to the amount of solute flowing out and the last term on the right hand side gives the solute transferred from the liquid phase into the adsorbent. The second term on the right hand side represents dispersion (or diffusion) in the bed. The concentration of solute (x) in the liquid is a function of bed length (z) and time (t). The solution of Eq. (5.54) is nonlinear and coupled, and so must be found numerically.

An approximate analysis for fixed bed could be performed using the break through curve. The exit concentration of the solute as a function of time remains zero, then it begins to rise abruptly, eventually reaching the feed concentration (x_F) (see Figure 5.23). The breakthrough concentration x_B, may be

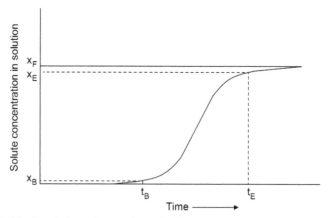

FIGURE 5.23   Break through curve in packed bed adsorber.

approximately 10% of x_F. The bed exhaustion concentration x_E, may be approximately 90% of x_E occurring at time t_E. The bed is judged ineffective after t_E.

The equilibrium zone where the bed is essentially saturated, is from $z = 0$ to $z = l(1-\Delta t/t_B)$, where Δt equals (t_E-t_B). The adsorption zone is $(l\Delta t/t_B)$ long where $l =$ bed length.

The equilibrium zone contains loaded adsorbent, that is,

$$q \text{ (equilibrium)} = q \, (x_F) \tag{5.55}$$

The adsorption zone contains half that much (this is an approximation)

$$q \text{ (adsorption)} = \tfrac{1}{2} \, q \, (x_F) \tag{5.56}$$

The fraction of the bed, which is loaded at break through is given by an approximate relation as:

$$\Theta = 1 - \Delta t/2t_B \tag{5.57}$$

The shorter the adsorption zone, the more completely will be the bed loaded.

5.10   PRECIPITATION

Precipitation (separation of a solid from a solution) can occur due to a chemical reaction and one of the products, which may be a solid, may drop out from the solution as a precipitate. The shape or size of the solid product in this unit operation is not as important as in the crystallization unit operation. Hence, fine crystals, power or agglomerates may form during this process. Agitation governs the size of the agglomerates. Agitation here helps in mixing of reactants, suspension of the solids formed and facilitate heat transfer to the wall. From lab experiments if it is confirmed that meso mixing has a strong effect on particle size then primary nucleation may be required (i.e., one may have to add seed crystals to initiate precipitation). If the particle size is related to the slurry concentration, then secondary nucleation may be taking place during the process. The solids that are formed may be amorphous or crystalline. An impure solid mixture could be dissolved in a solvent and the required solid could be precipitated out by manipulating the operating conditions leaving the other impurities dissolved in the solvent.

Salting out is a technique adopted to precipitate compounds of interest from the solution. In this process, another solute is added which decreases the solubility of the solute of interest and so precipitates it out of the solution. For example, when a solution of silver nitrate is mixed with a solution of sodium chloride, a reaction takes place to produce silver chloride, which precipitates out of the solution.

The precipitate can be separated by filtration, decanting, or centrifugation. The precipitation is generally carried out in a stirred tank. The reactions are operated in semi batch or continuous mode. In the former design, one of the reactants is taken in the vessel and the other reactant is added slowly so that the precipitated product settles down.

5.11   ABSORPTION

CO_2, Cl_2, CO, etc., from waste air or toxic gases from boilers are removed by absorbing them with water, alkali or monoethanolamine, etc. In gas liquid absorption, the gas enters the bulk phase of the liquid and gets dissolved or reacted (whereas adsorption is surface dependent). So absorption is a bulk phenomenon.

Henry's law, states that the concentration of a gas that dissolves in a liquid (C_A) is proportional to the partial pressure of the gas (p_A) over the liquid at low dilution.

$$p_A = K_H X_A = K_H C_A/C_T \tag{5.58}$$

where K_H is the Henry's law constant, C_T is the sum of concentration of all the species and X_A is the mole fraction of component A. Henry's law and Raoult's law, appear to be the same and both of them hold good for ideal systems.

Henry's law constant for a large number of gases are given in Sander (1999), (Compilation of Henry's Law Constants for Inorganic and Organic Species of Potential Importance in Environmental Chemistry (Version 3) – http://www.henrys-law.org).

Graham's Law states that when gases are dissolved in a liquid, the relative rate of diffusion of a gas is proportional to its solubility in the liquid and inversely proportional to the square root of its molecular mass.

Packed columns are used as absorbers (Figure 5.24), where the gas and liquid come in intimate contact. The contact could be co or countercurrent. Height of the column determines the removal efficiency while the diameter of the column determines the maximum gas throughput. In a countercurrent operation (Figure 5.24b), where the gas enters from the bottom, at a certain value of gas flow the liquid may stop descending and this phenomena is known as flooding. Flooding limits the maximum throughput through the absorber. Flooding depends on the pressure drop across the column, and column diameter.

In a dry column, the gas flow is turbulent. The pressure drop increases with gas flow ($\Delta P \propto v_G^{1.8-2.0}$) as well as increasing flow of liquid. Suddenly the pressure drop rises rapidly with gas flow and the liquid hold up in the column also rises. This is called the loading point (Figure 5.25). If the gas flow rate is increased, further flooding occurs. Here, the pressure drop rises drastically and the liquid flowing down stops. Generally, the column is operated at 60–65% of the flooding velocity. The type of packing used determines the interfacial area between the gas and liquid and the pressure drop.

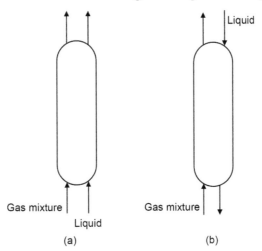

FIGURE 5.24 Continuous absorption column (a) cocurrent (b) countercurrent.

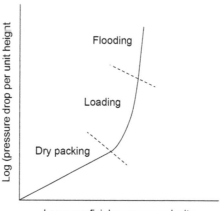

FIGURE 5.25   Loading and flooding point in packed bed.

The rate of gas transfer from the bulk gas to the liquid depends on the mass transfer coefficient. Several correlations are reported in the literature for this constant. Mass transfer coefficient depends on fluid and packing characteristics, tube diameter and gas flow. According to the two-film theory the total resistance for liquid-to-gas mass transfer is the sum of resistances contributed by the stagnant interfacial liquid and gas films. If the gas flow is turbulent or if the concentration of the solute gas is high then it is reasonable to ignore the resistant offered in the gas side.

For turbulent aeration and agitation, the equation for mass transfer coefficient is given by (Bailey and Ollis, 1977),

$$\frac{k_L D}{\partial} = Sh = 0.13 Sc^{1/3} Re^{3/4} \tag{5.59}$$

where D = agitator diameter; ∂ = oxygen diffusion coefficient (can be estimated from Wilke–Chang correlation (Wilke and Chang, 2004); k_L = gas to liquid mass transfer coefficient; Re = Reynolds number = $Du\rho/\mu$; Sc = Schmidt number = $\mu/\rho\partial$; Sh = Sherwood number is α $(P/V)^{1/4}$, where P/V is the power input by the agitator per unit volume of liquid; U = fluid velocity; ρ, μ = density and viscosity of the fluid, respectively.

In air sparged vessels without agitation when there are bubble swarms then for bubble diameter, d < 2.5 mm the equation for mass transfer coefficient is given by (Bailey and Ollis, 1977).

$$Sh = 0.31 Gr^{1/3} Sc^{1/3} \tag{5.60}$$

$$Gr= \text{Grashof number} = d\ \rho_g(\rho-\rho_g)g/\mu^2 \qquad (5.61)$$

where ρ_g = density of the gas, and d = gas bubble.

KEYWORDS

- **absorption**
- **adsorption**
- **dialysis**
- **isoelectric focusing**
- **liquid–liquid extraction**
- **membrane filtration**
- **pervaporation**
- **precipitation**
- **super critical extraction**

REFERENCES

1. Brown, G. G., Unit Operations, John Wiley & Sons, 1956.
2. Cusack, R. W., Glatz, D., et al. "A Fresh Look at Liquid-Liquid Extraction," Chemical Engineering, February, March & April 1991.
3. Firdaous, L., Saheb, T., Schlumpf, J-P., Maleriat, J-P., Bourseau, P., Jaouen, P. SIMS 2005, SIMS 2005 46th Conference on Simulation and Modeling, ISBN 82-519-2093-0. ISBN: 9788251920933.
4. Garcia, M. E. F., Habert, A. C., Nobrega, R., Piers, L. A. Use of PDMS and EVA membranes to remove ethanol during fermentation, Edited by R. Bakish. Proc. 5th Int. Conf. on Pervaporation Process in the Chemical Industry, Bakish Materials Corporation, Englewood, NJ, 1991, 319–330.
5. Hermans, P. H., Bred'ee, H. L. Principles of the mathematical treatment of constant-pressure filtration, J. Soc. Chem. Ind. 1936, 55T.
6. Ho, C.-C., Zydney, A. L., Effect of membrane morphology on system capacity during normal flow microfiltration, Biotech. Bioeng. 2003, 83(5), 537–543.
7. Kidney International, Editorial Review. Kinetic modeling of hemodialysis, hemofiltration, and hemodiafiltration. A mathematical model of multicomponent mass transfer in electrodialysis, 1983, Vol. 24, pp. 143–151.

8. Martin Chandler, Andrew Zydney, Effects of membrane pore geometry on fouling behavior during yeast cell microfiltration, Journal of Membrane Science 2006, 285, 334–342.

9. McCabe, W. L., Smith, J. C. Unit Operations of Chemical Engineering, McGraw-Hill, 1956.

10. Membrane Filtration Handbook Practical Tips and Hints, by Jørgen Wagner, BSc Chem. Eng., Second Edition, Revision 2. November 2001, © 2000, 2001 Printed by Osmonics, Inc.

11. Michael D., LaGrega, Phillip L. Buckingham and Jeffery C. Evan. Hazardous Waste Management, McGraw Hill, Inc., 1994.

12. Perry, R. H., Green, D. Perry's Chemical Engineers' Handbook, 6th Edition, McGraw-Hill, 1988.

13. Robbins, Chem. Eng. Prog., 1980, 76(10), 58–61.

14. Schafer, A. I., Fane, A. G., Waite, T. D. cost factors and chemical pretreatment effects in the membrane filtration of waters containing natural organic matter, Wat. Res. 2001, Vol. 35, No. 6, 1509–1517.

15. Seungkwan Hong, Ron S. Faibish, Menachem Elimelech. Kinetics of Permeate Flux Decline in Cross-flow Membrane Filtration of Colloidal Suspensions. Journal of Colloid and Interface Science 1997, 196, 267–277.

16. Song, L., Elimelech, M., DPc pressure drop across the cake layer, *J. Chem. Soc., Faraday Trans.* 1995, 91, 3389.

17. Xianshe Feng, Robert Y. M. Huang, Liquid Separation by Membrane Pervaporation: A Review. Ind. Eng. Chem. Res., 1997, 36(4), 1048–1066.

PROBLEMS

1. Derive the equation for overall separation efficiency for a countercurrent extraction system consisting of 'n' stages. How is it better than a crosscurrent operation consisting of 'n' stages

2. Derive the equation for the exit concentration of the solute as function of time for a continuous stirred adsorption unit when the solute is injected in the feed at time = 0 (the adsorption follows a linear relationship).

3. Derive the equation for the exit concentration of the solute as function of time for a packed bed system assuming n tanks in series when the solute is injected in the feed at time = 0 (the adsorption follows a linear relationship).

4. Calculate the flux across polyamide-urea membrane for a pressure of 30 bar, assuming the solids is completely retained or they are completely passed.

5. Estimate the electrodialysis power for a load of 10000 ohms, when the solution normality is 2, flow rate 10 m³/s, removal and current efficiencies of 65 and 75%, respectively, and number of cells is 4.

6. Estimate the fraction of the adsorption bed, which is loaded if the breakthrough time is 2 h and exhaustion time is 4 h.

7. In hemodialysis the concentration of the impurity in the inlet of the dialyzer is 10 mM. The blood flow rate is 100 mL/min. If the dialyzer clearance is 0.5 mL/min

estimate the concentration of the impurity in the outlet. If there is a 5% escape of blood fluid then estimate the concentration of the impurity in the outlet.

8. A gas containing 2% by volume CO_2 is passed through water. Estimate the maximum CO_2 that will dissolve at 25°C.

9. In concentration polarization if bulk and surface concentration of the solute is 5 and 2 mM, respectively, then estimate the flux through the membrane if mass transfer coefficient is 0.01 mol/cm²/min.

10. Estimate the number of stages for a countercurrent extractor train, if the $E_F = 0.5$ and the extraction efficiency is (a) 90% and (b) 99%.

11. Derive the equation for Langmuir adsorption isotherm.

12. A solution of sucrose in water is concentrated by RO. It is found that, with a differential applied pressure of 6000 kPa, the rate of movement of the water molecules through the membrane is 25 kg m⁻² h⁻¹ for a 10% solution of sucrose. Estimate the flow rate through the membrane for a differential pressure of 12,000 kPa at 25°C.

13. If the sucrose concentration is 20% what will be the flow rate at the higher pressure in the previous problem.

14. It is desired to increase the protein concentration in whey by a factor of 10 by the use of membrane filtration to give an enriched fraction to produce a 50% protein whey powder. The whey initially contains 5% of total solids, 10% of these being protein. Pilot scale measurements on this whey show that a permeate flow of 40 kg m⁻² h⁻¹ can be expected, so that the plant is required to handle 40,000 kg in 8 h. Estimate the number of membrane sheets of 10 m² area required.

15. Assume that the membrane rejection of the protein is over 99% and calculate the membrane rejection of the non-protein constituents in the previous problem.

CHAPTER 6

PRODUCT ENRICHMENT

CONTENTS

This chapter deals with product enrichment and the two unit operations are chromatography and distillation. Chromatography is a downstream processing technique, which can be used for concentrating or polishing the desired product. Hence, it can be used in the intermediate or final stages of the flow sheet. It is well suited for temperature sensitive materials including proteins, enzymes, peptide, etc. Distillation is well suited for recovering pure products, which are not temperature sensitive. The other separation process discussed in the Chapter 5 (such as, precipitation, ultra-filtration, dialysis, etc.) can also be used for final purification depending upon the requirements.

6.1 CHROMATOGRAPHY

Industrial scale chromatographs are still not fully understood in large scale and appears to be an art. Two points to keep in mind are (i) compare the product and its most closest impurity, and (ii) consider the scale at which the purification is to be operated. Some methods are more suited at milligrams per day scale and others are good at tons per hour. Electrophoresis is another purification technique, which is not practiced in large scale.

Chromatographic techniques can be divided into two broad categories namely, non-adsorptive and adsorptive methods. If the solute interacts with the chromatographic matrix then it is called adsorptive. Example of the non-adsorptive method is gel-filtration chromatography, which separates molecules on the basis of their size. It is a sieving technique where the molecules being fractionated do not adsorb onto the matrix or interact with it. All other chromatographic methods including ion-exchange chromatography; affinity chromatography; reverse- and normal-phase chromatography; and hydrophobic-interaction chromatography are adsorptive methods. The principle of separation here is due to the interaction between the solid matrix and the solute molecules. The interaction force could be varying from weak to strong.

Chromatography technique adds to the final product price. Technique such as, affinity chromatography is an expensive method and it is used to obtain extremely high purity product. The stationary phase is a liquid or a solid and is fixed in the system. The other is the mobile phase, a fluid, which is flowing through the chromatographic system. In gas chromatography the mobile phase is a gas and in liquid chromatography the mobile phase is a liquid. Table 6.1 lists the type of chromatography and the physical principle. The solute has different distribution/partition between the two immiscible phases.

TABLE 6.1   Principle Involved in the Chromatography Separation

	Physical property	**Type of chromatography**
Polarity	Volatility	Gas–liquid
		Liquid–liquid
		Liquid–solid
Ionic	Charge	Ion–exchange
Hydrophobic	Hydrophobicity	Hydrophobic interaction/reversed-phase
Size	Diffusion	Gel permeation/size exclusion
Shape	Molecular recognition	Affinity

A pulse of solute is injected at the entrance of the packed column while the solvent flow is kept constant and continuous. After certain time the pulse comes out from the other end of the column. The pulse will have entered as a sharp narrow, concentrated peak, but it will exit dispersed (broad) and diluted (Figure 6.1). Broadening of peak happens due to diffusion. Diffusion coefficient measures the rate at which a compound moves randomly from a region of high concentration to a region of lower concentration. Pulses of different solutes all leave the bed diluted, but they leave at different times, as shown schematically in Figure 6.2. The solute that is

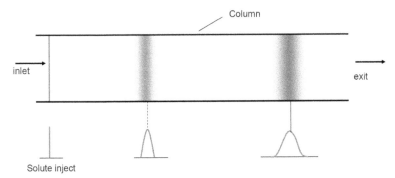

FIGURE 6.1   Concentration profile as it travels through the column.

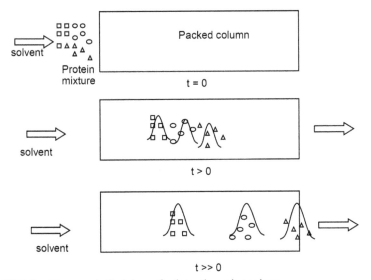

FIGURE 6.2   Movement of mixture of solutes through a column.

not adsorbed much leaves the column quickly. The solute most strongly adsorbed leaves slowly.

The peak exiting the chromatographic column under ideal conditions appears as a Normal or Gaussian distribution (Figure 6.3). The sample that is introduced into the column is infinitely narrow and is called δ- or Dirac delta function and its peak width is negligible.

The Gaussian curve or shape of an output from a chromatographic column is given by the following relation

$$c = c_0 \exp\left(-\frac{[t/t_0 - 1]^2}{2\sigma^2}\right)$$ (6.1)

where c_0 is the maximum concentration, t_0 is the time at which this concentration exits, and $t_0\sigma$ is the standard deviation of the peak (Figure 6.4). Taking ln on the both sides of this equation leads to

$$(t/t_0 - 1)^2 = 2\,\sigma^2\,\ln(c_0/c)$$ (6.2)

where σ can be estimated by plotting $(t/t_0-1)^2$ versus $2\ln(c_0/c)$; since the slope is σ^2. The data from the ascending and descending portions of the profile will fall along the same line.

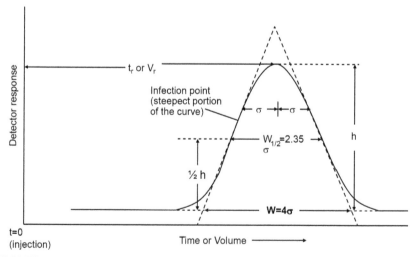

FIGURE 6.3   Solute moving through a column spreads into a Gaussian shape.

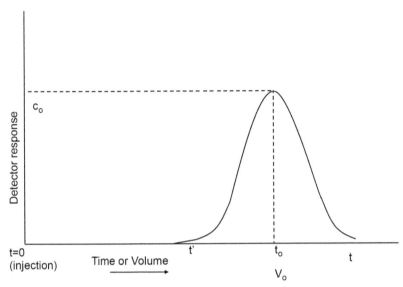

FIGURE 6.4   A typical chromatogram.

A similar equation in terms of the volume eluted V can be written as

$$c = c_0 \exp\left(-\frac{[V/V_0 - 1]^2}{2\sigma^2}\right) \tag{6.3}$$

Where V_0 is the volume required to elute the maximum concentration c_0 and $V_0\sigma$ is now the standard deviation. Note that V equals Qt. σ can be found by plotting $(t/t_0 - 1)^2$ versus $2\ln(V_0/V)$; since the slope is σ^2.

A mixture may have j solutes and concentration of a solute i eluted out will be c_i, which will be a function of time. The eluted solutes are collected as fractions, some of which are enriched with a particular solute.

The total amount of solute eluted between two times, t' and t (Figure 6.4) is given by

$$\text{Amount of } i \text{ eluted in this time period} = \int_{t'}^{t} cQdt \tag{6.4}$$

where c is solute concentration and Q is the solvent flow. The total solute in the column is obtained by integrating from time 0 to ∞.

$$\text{Total of i-th solute in the column} = \int_0^\infty cQdt \qquad (6.5)$$

If the i-th solute eluting is collected between time t' and t, then the yield of i-th solute is the ratio of these integrals.

$$\text{Yield of i-th} = \frac{\int_{t'}^t cQdt}{\int_0^\infty cQdt} \qquad (6.6)$$

By substituting the equation for Gaussian distribution curve (eqn. 6.1) into the equation (6.4) one gets

$$\text{Amount eluted} = \int_{t'}^t Qc_0 \exp\left(-\left[\frac{t/t_0 - 1}{2\sigma^2}\right]^2\right) dt \qquad (6.7)$$

Equation (6.6) for yield will get modified to

$$\text{Yield of i-th} = \frac{1}{2}\left\{ erf\left[\frac{t/t_0 - 1}{\sqrt{2}\sigma}\right] - erf\left[\frac{t'/t_0 - 1}{\sqrt{2}\sigma}\right]\right\} \qquad (6.8)$$

The analogous equation in terms of elution volume will be

$$\text{Yield} = \frac{1}{2}\left\{ erf\left[\frac{V/V_0 - 1}{\sqrt{2}\sigma}\right] - erf\left[\frac{V'/V_0 - 1}{\sqrt{2}\sigma}\right]\right\} \qquad (6.9)$$

erf(x) is the error function and is defined as

$$erf(x) = \frac{\sqrt{2}}{\pi}\int_0^x e^{-u^2} du \qquad (6.10)$$

If t and t' are equal, then there is no yield. If $t'=0$, (i.e., if one starts collecting sample from zero time) the error function containing t' will be equal to -1, and the equation for yield becomes

$$\text{Yield} = \frac{1}{2}\left\{1 + erf\left(\frac{t/t_0 - 1}{\sqrt{2}\sigma}\right)\right\} \qquad (6.11)$$

When, $t = t_0$, half the pulse has been eluted, and the yield is 50%. ($erf(0) = 0$).
The purity is weighted by the amounts present and the denominator will contain all the solutes

$$\text{Purity of solute } i = \frac{c_0(i)\,yield(i)}{\sum_j c_0(j)\,yield(j)} \qquad (6.12)$$

As before, the summation is over all the j solutes in the system. If sample were collected between time t_1 and t_2, one would get predominantly compound 1 and part of compound 2 (Figure 6.5). Similarly, if one collects sample between time t_2 and t_3, one would get predominantly compound 2 and part of compound 1. If one collects sample between time t_4 and t_5, one would get pure compound 3.

There are four processes that occur during the movement of the sample through the chromatographic column and they are:

1. the solute is transferred from the bulk of the solution to the surface of the solid stationary phase.
2. from the surface it diffuses into the solid matrix.

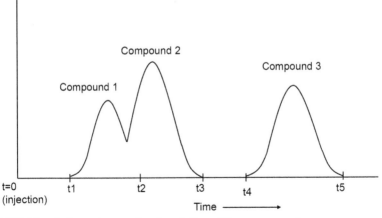

FIGURE 6.5 A chromatogram showing elution of three compounds.

3. it interacts reversibly with the packing; this interaction may include adsorption and other forces and eventually desorption. (If there is an irreversible reaction then the matrix cannot be used again).
4. the desorbed solute diffuses back out of the packing to the solid surface.
5. it diffuses from the surface back into the bulk of the solution.

6.1.1 ION EXCHANGE

Ion exchange chromatography (IEC) separates proteins based on their charge. The stationary phase will contain ligands of certain charge and proteins in the mixture of the opposite charge preferentially bind to it. Proteins with the same charge of that of the ligands or uncharged proteins elute out first (Figure 6.6).

There are two types of ion exchange systems, namely the cation and the anion. Immunoglobulin G (IgG) and Bovine Serum Albumin (BSA)

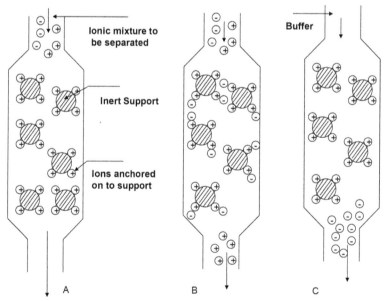

FIGURE 6.6   Ion exchange chromatography (A – start up, B – separation, C – regeneration).

are separated on a strong anion exchanger using a linear salt gradient. It is possible to either to bind the proteins and ions of interest and allow the contaminants to pass through the column, or to bind the contaminants and allow the molecules of interest to pass through. Generally, the former is more useful since it allows a greater degree of fractionation and concentrates the compound of interest. But this approach may overload the column. The latter is preferred if the concentration of the contaminants are less in the mixture. Figure 6.6 shows the three steps involved in the IEC. An ion exchanger consists of solid matrix to which charged groups are covalently bound. These charged groups are associated with the counter-ions in the mobile phase. These counter-ions can be reversibly exchanged with other ions of the same charge without altering the matrix. Systems are also designed to have both anions and cations in the stationary phase. This design is useful if one is interested in removing just salts from a protein solution.

Functional groups used on ion exchanger matrix vary depending upon whether it is an anion or cation. For an anion exchanger, functional groups that are used include, cellulose or agarose; diethylamino-ethyl; quaternary aminoethyl; and quaternary ammonium. For cation exchanger's functional groups that are used include, carboxymethyl; sulphopropyl; and methyl sulphonate. Sulphonic and quaternary amino groups form strong, while other groups form weak ion exchangers. The variation of ionization as a function of pH, determines the strength of the ion exchangers and not on the strength of binding between the protein and the ligand.

A good ion exchanger should have the following properties:

1. Capacity of loading of the sample should not change with change in pH due to loss of charge from the ion exchanger.
2. Interaction between the ion exchanger and the solute should be based on simple mechanism.
3. Ease of scale-up, that is, the data obtained from electrophoretic titration curve should be used to design ion exchange system.

The capacity of IEC is affected by (i) pH, (ii) ionic strength of the buffer, (iii) nature of the counter-ion, (iv) flow rate of the solvent, and (v) temperature. Increasing flow rate decreases the dynamic capacity.

6.1.2 HYDROPHOBIC INTERACTION

Hydrophobic interaction chromatography (HIC) separates proteins on the basis of the differences in the hydrophobic interaction strengths between the proteins and the stationary phase containing the immobilized hydrophobic groups. All proteins will have hydrophobic groups such as, hydrophobic amino acids (i.e., those with nonpolar R groups—alanine, phenylalanine, valine, tryptophan, leucine, isoleucine, and methionine).

These amino acids are generally buried inside the protein molecule in the aqueous environment. Some non-polar amino acids are located on the surface of the molecule as distinct hydrophobic regions due to folding constraints. The difference in the hydrophobicity of one protein with respect to another, forces it to selectively get adsorbed and eluted out at a different rate with respect to the other protein from the hydrophobic matrix.

The HIC process involves (Figure 6.7), (i) loading a protein mixture suspended in a high salt solution onto a matrix containing the hydrophobic ligands, then (ii) eluting the proteins by either: (a) decreasing the salt in the solution, or (b) changing the polarity of the water phase by adding nonionic detergents or organic solvents.

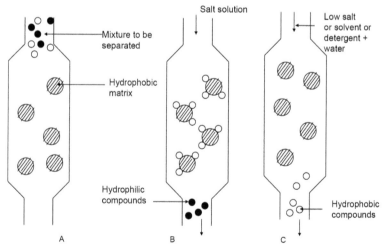

FIGURE 6.7   Hydrophobic interaction chromatography (A – start up, B – separation, C – regeneration).

HIC is generally practiced at the early stages of the separation process, when protein is precipitated in the presence of salt. HIC involve use of high salt concentrations; hence, this method is ideal if the prior step also uses salts. HIC, IEC, and gel filtration are useful techniques to remove protein without affecting its activity.

The important parameters to be considered while selecting a hydrophobic interaction chromatography matrix are:

1. type of ligand;
2. coverage of the ligand in the matrix;
3. composition of the matrix support;
4. type of salt used in the buffer;
5. salt concentration;
6. pH and temperature; and
7. additives used in the buffer.

The type of ligand immobilized on the stationary matrix determines the selectivity of the HIC. Two types of ligand are immobilized on HIC—alkyl-chain of different lengths and aryl groups (usually phenyl). The former shows true hydrophobic nature. The protein-binding capacities of HIC matrices increase with increasing the hydrocarbon chain length. Two common matrices used in HIC are 4% and 6% cross-linked agarose. They are strong hydrophilic carbohydrates. Smaller particle size leads to higher resolution. Matrices with particle sizes of 34 μm are considered as high-resolution media. Smaller the particle size higher the pressure drop in the column and it is inversely proportional to the square of the particle size of the packing. Addition of salts leads to "salting out" effect. This facilitates the interaction between the immobilized ligand and the protein. As the concentration of salt is increased, the amount of protein bound to the immobilized ligand also increases linearly.

Certain salts promote interaction and certain others promote elution of the protein from the matrix. The Hofmeister series of anions and cations are shown in Table 6.2. As we move from right to left for anions "salting out" (also known as structure forming) effects are increased. As we move from left to right for anions "salting in"effect (chatotrophic) is increased. Similarly, as we move from left to right for cations, "salting in" effect is increased so as we move from right to left for cations, "salting out" effect is increased.

TABLE 6.2   Hofmeister series describing the effect of anions and cations on protein precipitation

$PO_4^{3-} > SO_4^{2-} > CH_3COO^- > Cl^- > Br^- > NO_3^- > ClO_4^- > I^- > SCN^-$
←Increasing salting out effect
$NH^{4+} < Rb^+ < K^+ < Na^+ < Cs^+ < Li^+ < Mg^{2+} < Ca^{2+} < Ba^{2+}$
Increasing salting in effect →

Changing pH improves interaction between the ligand and the proteins. So proteins that do not bind to a hydrophobic-interaction stationary matrix at neutral pH, will bind at acidic pH to it.

Desorption of compounds bound to an immobilized hydrophobic ligand is the second step and can be achieved by:

1. decreasing the concentration of salt,
2. adding an organic solvent to the elution buffer, or
3. adding neutral detergent to the elution buffer.

The first method is very common and it can be done in linear or step-wise fashion. Using water as an eluent also leads to desorption. Adding low concentrations of water-miscible alcohols, detergents, and salting-in salts (i.e., chaotropic salts that decrease precipitation of hydrophobic compounds), leads to the weakening of the protein-ligand interaction, and hence, leads to the desorption of the bound protein.

In continuous gradients, salt concentration is increased or decreased linearly and continuously. In step-gradient, one or more salt solutions of a discrete concentration are passed through a column. Step-wise elution is preferred for large-scale applications because it is simpler and more reproducible than the linear elution.

6.1.3   REVERSED-PHASE CHROMATOGRAPHY

Normal-phase (polar) chromatography uses polar stationary phase such as, silica gel and nonpolar solvents such as, hexane (Figure 6.8). Here, there is an interaction between the polar functional groups in the protein and the polar groups on the stationary matrix. Low polarity substances are eluted first, followed by increasing polarity compounds. The elution sequence for normal silica gel column starts from carboxylic acids,

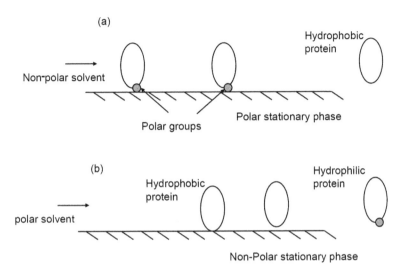

FIGURE 6.8 (a) Normal-phase (polar) chromatography and (b) reversed-phase chromatography.

amines, sulfones, sulfoxides, alcohols, amines, esters, aldehydes, ketones, nitro compounds, ethers, sulfides, aromatics, organic halogen compounds, olefins and finally, alkanes.

In "reversed-phase" chromatography, as the name implies the stationary phase is nonpolar and elution is carried out with polar solvents. So polar proteins elute out first and non-polar ones elute at the end (Figure 6.8). These stationary phases are produced by attaching non-polar groups such as, C-18 on the silica gel. Reversed-phase materials are more expensive than normal stationary phases.

Surfactants are used in the mobile phase of HIC because of their advantages which include, (i) their ability to solubilize hydrophobic compounds; (ii) selectively partition many solutes into micelles; (iii) low cost; and (iv) the ability to change the polarity of the micellar mobile phase. This can be achieved by changing the concentration of surfactant in the solution. Structurally similar solutes can be separated with a micellar mobile phase because micelles solubilize and bind a variety of solute molecules via non-bonded interactions (such as, hydrophobic, electrostatic, and hydrogen-bonding).

There are several critical parameters in reversed-phase chromatography that affect the performance namely, column length, temperature, solvent used, ion suppression, and ion pairing agents. Many other points that need to be kept in mind are: (i) Higher molecular weight biomolecules such as, ,proteins, large peptides and nucleic acids can be purified on short column. Increasing column length improves the resolution only marginally. The resolution of small peptides may improve by increasing column length. (ii) Resolution of larger biomolecules is insensitive to flow rate. Flow rate in long columns, decreases resolution due to increased longitudinal diffusion of the solute molecules as they travel longer along the length of the column. (iii) For low molecular weight solutes such as, short peptides and oligonucleotides temperature has a major effect. (iv) Viscosity of the mobile phase used in reversed-phase chromatography decreases with increasing column temperature. Since, mass transport of solute between the mobile phase and the stationary phase is a diffusion-controlled process, decreasing solvent viscosity may improve mass transfer coefficient and hence, lead to higher resolution.

6.1.4   ADSORPTION CHROMATOGRAPHY

Absorption chromatography is used for non-polar or mildly polar organic molecules. The stationary phase is made up of polar solid such as, silica or alumina beads. The analyte competes with the mobile phase for the active sites on the surface of the matrix. Attraction of the analyte to this matrix is due to adsorption forces. Organic solvent including hexane or other hydrocarbons act as the mobile phase and its composition can be modified to affect the partition coefficient of the analytes in the mixture. Isomeric mixtures can be resolved by adsorption chromatography.

6.1.5   LIQUID–LIQUID PARTITION CHROMATOGRAPHY

In liquid–liquid partition chromatography as the name implies there are two liquids. The inert support or stationary matrix is coated with a polymeric layer or with a liquid that is insoluble in the mobile phase. The separation is based on the relative solubility or partitioning of the solute in the mobile liquid phase and in the stationary phase (Figure 6.9) and the relation is given by

$$K = x/y \qquad (6.13)$$

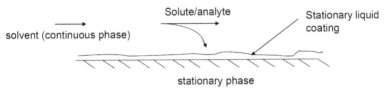

FIGURE 6.9   Liquid–liquid partition chromatography.

where K is the partitioning coefficient and x and y are the concentration of the analyte or solute in the stationary and continuous/mobile phases, respectively. There are two types of partition chromatography (similar to solid stationary phase), namely, (i) normal-phase, and (ii) reversed-phase. Normal-phase chromatography makes use of highly polar stationary phase and hydrophobic solvents such as, hexane for the mobile phase. The least polar component is eluted first and polar solutes much later. An increase in the polarity of the mobile phase will decrease the retention time. Reversed-phase chromatography is used to separate highly polar analytes. In conventional chromatography such systems gives long retention times and peak tailing. In reversed-phase a non-polar stationary phase such as, a hydrocarbon is used as the coating liquid while a relatively polar solvent (such as, alcohols) is used as the mobile phase.

6.1.6 GEL PERMEATION CHROMATOGRAPHY (GPC) OR SIZE EXCLUSION CHROMATOGRAPHY

This chromatography (GPC) differs from the other ones in the sense that there are no chemical or physical interaction between the analyte (solute) and the stationary phase. It has a column that is packed with a porous material. Small molecules (which have size smaller than the pores) will diffuse into the pores, while larger ones will not. So the latter will pass through the column more rapidly and exit or leave first (Figure 6.10). The elution time is inversely proportional to the size (or molecular weight). The stationary matrix for lipids is based on copolymers of styrene and divinylbenzene with pore sizes in the range 50–500Å. Spherical particles of 5–10 µm in diameter are preferred here. Columns of different pore sizes are used in series to improve separation efficiency. The mobile phases

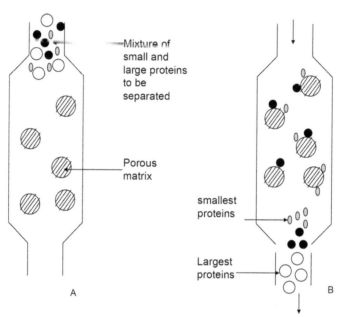

FIGURE 6.10   Size exclusion chromatography (A – start up, B – Separation).

include tetrahydrofuran, dichloromethane and toluene. Aqueous mobile phases are preferred for the separation of lipoproteins and other biomolecules. Detectors that are used here include viscometry, low angle laser light scattering, refractive index, ultraviolet/visible absorbance, fluorescence, and precision differential refractometry coupled with ultraviolet detection.

6.1.7   AFFINITY CHROMATOGRAPHY

Affinity chromatography is based on the principle that every biomolecule recognizes another target complementary molecule. It is based on binding and interaction; and it is used exclusively for proteins and antibodies. The binding is between a ligand and a protein or enzyme (6.11). Either one of them is immobilized on a polymeric matrix through covalent bonding. The cell extract is passed through the column and the counterpart is selectively captured (Figure 6.11). All proteins with poor affinity for the bound inhibitor will pass directly through the column, whereas the one

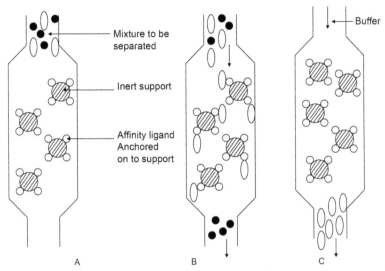

FIGURE 6.11   Affinity chromatography (A – start up, B – Separation, C – regeneration).

that is complementary to the inhibitor will be retarded in proportion to its affinity constant. A spacer arm is used to maintain some distance between the ligand and the support matrix to improve the accessibility of the latter (Figure 6.12). This system forms a very selective stationary phase that will only bind to a protein of the ligand pair. Affinity chromatography is the only technique that enables the purification of a biomolecule on the basis

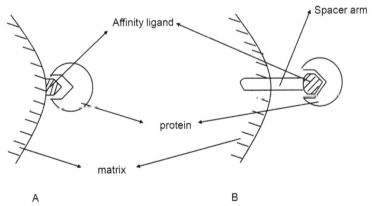

FIGURE 6.12   Spacer arm – used to ensure full accessibility of the affinity ligand (A – poor, B – ideal design).

of its biological function or individual chemical structure. The technique can be used to separate· (i) biomolecules from denatured or functionally different forms, (ii) isolate pure substances present at low concentration, or (iii) to remove specific contaminants.

A biological molecule interacts with another molecule through binding at specific sites. Examples are: (i) binding of an antigen to an antibody, (ii) a substrate, inhibitor, or co-factor to an enzyme, (iii) a regulatory protein to a cell surface receptor. The binding forces between ligand and protein include ionic, electrostatic, hydrogen bond and hydrophobic interactions. The shape and size of the groups in the two binding molecules or ligands will be unique and the two will fit together like a lock and key,

Any component can be used as a ligand to purify its compliment-binding partner. Some biological interactions, frequently used in affinity chromatography, are

1. Enzyme – substrate analog, inhibitor, cofactor;
2. Antibody – antigen, virus, cell;
3. Lectin – polysaccharide, glycoprotein, cell surface receptor, cell;
4. Nucleic acid – complementary base sequence, histones, nucleic acid polymerase, nucleic acid binding protein;
5. Hormone, vitamin – receptor, carrier protein;
6. Glutathione – glutathione-S-transferase or GST fusion proteins;
7. Metal ions – Poly (His) fusion proteins, native proteins with histidine, cysteine; and/or
8. Tryptophan residues on their surfaces.

The desired target (which is captured) is eluted out of the column by changing the external conditions, which may include ionic strength, pH, solvent or temperature (Figure 6.13). Best techniques used to disrupt the ligand – protein interaction are:

(i) change to acidic conditions (to pH 2–4) – generally used for protein and for antibody ligand affinity;
(ii) Increase ionic strength (e.g., heparin); and
(iii) use specific eluents such as, the immobilized ligand (Figure 6.14) or an analog in free solution (Figure 6.15).

Interaction between the target molecule and the affinity ligand is reversible and has an equilibrium dissociation constant (K_D).

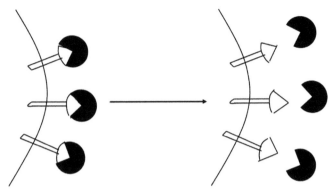

FIGURE 6.13 Elution (desorption) of ligand bound protein by changing environment (KD values suitable for elution are typically in the range of 10^{-1}–10^{-2}).

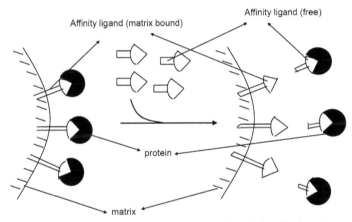

FIGURE 6.14 Elution by displacement (a free ligand is added to displace the target from the matrix-bound ligand).

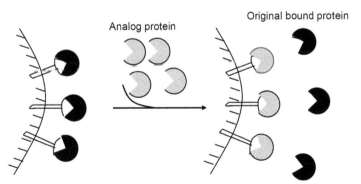

FIGURE 6.15 Addition of analog in free form replaces the bound protein (e.g., Elution of HIS tagged proteins from HiTrap chelating by adding imidazole).

$$L + P \leftrightarrow LP \qquad (6.14)$$

$$K_D = [L]\,[P]/[LP] \qquad (6.15)$$

The smaller the value of K_D the stronger is the binding. Values for good binding are typically in the range of 10^{-4}–10^{-6} M and values greater than 10^{-4} leads to weak binding and the molecule may "leak" as a dilute broad zone and get washed. K_D value can be modified by changing conditions as mentioned before.

If a ligand binds too strongly to the target then it will be difficult to elute it out without resorting to harsh conditions. Such conditions may destroy the biological activity or force it to irreversibly get adsorbed to the ligand. Ideal elution requires the target molecule to desorb completely from the affinity ligand in elution buffer so that the system is fully regenerated and the target molecule is completely recovered. If binding or desorption is slow or poor then one need to try another affinity system.

Leakage of target molecule during sample loading indicates that the sample residence time is too short for complete binding (so higher residence times are required). A way to extend the residence time is to inject the sample in small portions and stop the flow after each injection. The duration of the flow stop has to be determined by trial and error.

Immunoaffinity chromatography is a modification to affinity chromatography, which is used in antibody columns to purify antigens; for isolating receptors, enzymes and DNA fragments; for removal of toxic components from blood by hemoperfusion; and for large-scale preparations of monoclonal antibodies. Antibodies can be produced against any compound with ease, and these can be used to purify the parent compound by immobilizing them on a stationary matrix. Affinity technique is an expensive method, which adds to the cost of purification.

Another affinity chromatography method involves incorporating various temporary affinity tags or affinity tails including His-Tag to the protein, which is then purified on metal-chelate (such as, Ni) affinity chromatography column. The affinity tag can therefore be used to purify a fusion protein on an immunoaffinity column, and the native protein can be recovered by cleavage using enterokinase. The fusion proteins are produced with large proteins such as, glutathione transferase, protein A, maltose-binding proteins, cellulose binding domains and biotinylated sequences. Each one of these will have its own specific ligand through which the fused protein can be purified.

6.1.8 ION-INTERACTION CHROMATOGRAPHY

This is a technique for performing the reversed-phase chromatography of charged solutes. Hydrophilic ionic solutes (low-molecular-weight compounds such as, hydrophilic amino acids, di- and tripeptides and zwitterionic solutes) are not retained on lipophilic stationary matrices when reverse-phase eluents are used. Such solutes can be separated by the addition of a lipophilic reagent ion with the opposite charge to the mobile phase. Ion-interaction reagents include salts that consist of a large lipophilic organic ion and an inorganic counter ion such as, sodium sulfonate or tetraalkylammonium chloride. This technique is known as ion-interaction chromatography. A combination of electrostatic and adsorptive forces leads to the separation (Figure 6.16). The lipophilic ion added forms a dynamic equilibrium between the eluent and the stationary phase such that an electric double layer is formed at the surface. This adsorbed ions form a primary layer of charge. A secondary layer of counter ions are formed on top of this. A solute with an opposite charge to the original ion (complementary ion) competes for a place in the secondary charged layer and then moves into the primary layer. This leads to a decrease in the total charge of the layer. So another ion moves into the primary layer to maintain charge balance. So the use of ion pairing modifiers in the mobile phase allows the

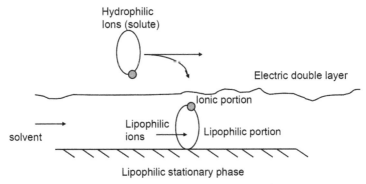

FIGURE 6.16   Ion-interaction chromatography for reversed-phase systems.

use of reversed-phase chromatography method for charged solutes, especially for deprotected oligonucleotides and hydrophilic peptides.

6.1.9   ION-SUPPRESSION REVERSED-PHASE CHROMATOGRAPHY

This technique is preferred when stationary matrix/phase is non-polar and the mobile/continuous phase is polar but under the operating conditions may become non-polar. A pH buffer is added to the mobile phase to suppress the ionization of the analyte. Then polar analytes will elute out first. Here, the ionizable compounds must be kept in their ion-suppressed form. Ionized compounds are highly polar, which under reversed-phase conditions would elute very quickly from the column leading to poor separation from other ionized analytes. By using buffer and at appropriate pH the compounds are forced into their ion suppressed forms, so weak acids and bases may be effectively chromatographed under reversed-phase conditions. The pKa of a molecule can be used to determine the correct pH at which the mobile phase needs to be buffered. For organic acids such as, ethanoic acid pKa = 4.75 so the pH can be adjusted below its pKa. For organic bases such as, trimethylamine pKa = 4.19 so the pH is adjusted above its pKa. By this adjustment these weak organic acids and organic bases can be effectively separated in a reversed-phase chromatography.

6.1.10   THIN-LAYER CHROMATOGRAPHY

Thin-layer chromatography (TLC) is the most useful and simplest of all technique for the analysis of organic molecules because of (i) low cost, (ii) minimal sample clean up, (iii) wide choice of mobile phases, (iv) flexibility (v) easy sample detection, (vi) high sample-loading capacity, and (vii) ease of handling. TLC can be used for the qualitative analysis of complex biologic lipid mixtures as well as for quantification of wide range of organic compounds. Preparative TLC can be used to collect large quantities of pure compound for further analysis.

A solid adsorbent is coated onto a flat surface support such as, glass, plastic or aluminum as a thin (=0.25 mm) layer, sometimes mixed with a small amount of binder. The mixture to be separated is dissolved in a solvent and it is spotted onto this surface near the bottom (Figure 6.17). A solvent or a mixture of solvents is made to flow up the plate due to capillary action by dipping the bottom of this plate in this solution. Based on the partition coefficient, the solid will adsorb a certain fraction of each component of the mixture and the remainder will be eluted with the solvent solution. A substance that is strongly adsorbed by the solid will have a greater fraction of it, and thus, will spend more time on the support and less time eluting with the solution. A weakly adsorbed substance will have a smaller fraction of it adsorbed, and hence, will move up fast. Weakly adsorbed substance will move up the plate and strongly adsorbed substance will stay near the bottom of the plate.

Alumina, is the strongest adsorbent, followed by charcoal, and florisil, MgO/SiO_2 (anhydrous). Silica gel, is the least strong adsorbent. With alumina as the adsorbent, the solvents with least eluting power are petroleum ether (hexane; pentane) cyclohexane, carbon tetrachloride, benzene, dichloromethane, chloroform, ether (anhydrous), ethyl acetate (anhydrous), acetone (anhydrous), ethanol, methanol, water and pyridine. With alumina as adsorbent, the solvent with greatest eluting power is organic acids. Most strongly adsorbed are acids and bases (amines) while least strongly adsorbed are saturated hydrocarbons; alkyl halides, unsaturated hydrocarbons; alkenyl halides, aromatic hydrocarbons; aryl halides, polyhalogenated hydrocarbons, ethers, esters, aldehydes, ketones, and alcohols.

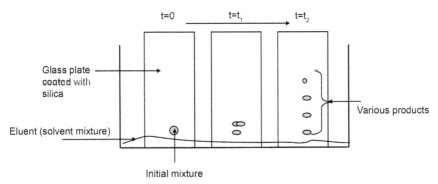

FIGURE 6.17   Thin layer chromatography.

The relationship between the distance traveled by the solvent and the compound is expressed as the R_f value, and the equation defining this given as

$$R_f \ value = \frac{distance\ traveled\ by\ the\ compound/}{distance\ traveled\ by\ the\ solvent} \qquad (6.16)$$

6.1.11 THEORETICAL PLATES/COLUMN EFFICIENCY

A theoretical plate is an ideal unit within which the eluent/solute in the continuous phase and in the stationary phase interact and reach an equilibrium after coming in close contact with each other. The long chromatographic column is assumed to be made up of many theoretical plates, an stages or units.

The following equations are used to calculate (see Figure 6.18) the number of theoretical plates (N) in a chromatographic column

$$N = 5.54(V_r/W_{1/2})^2 \qquad (6.17)$$

$$N = 5.54 \ (t_r/W_{1/2})^2 \qquad (6.18)$$

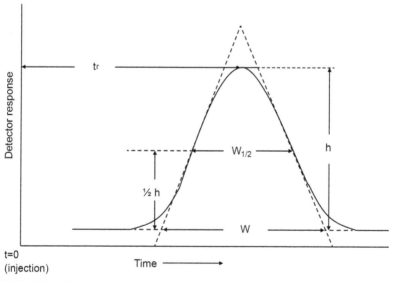

FIGURE 6.18   Chromatogram (detector response).

where V_r is the volume eluted from the start of sample injection to the peak maximum, $W_{1/2}$ is the peak width measured at half peak height. $W_{1/2}$ in the above two equations will have units corresponding to the numerator, so that the ratio will become dimensionless. t_r is the time at which the maximum occurs (Figure 6.18). Number of theoretical plates can also be calculated from peak width measured at the base (W) as

$$N = 16\,(t_r/W)^2 \tag{6.19}$$

The height of a theoretical plate is given as

$$H = L/N \tag{6.20}$$

where L is the length of the packed chromatographic bed. Ideal value for H is about two to three times the mean diameter of the packing beads. Thus, a good H value for a 90-μm bead is between 0.018 and 0.027 cm and a good H value for 34-μm matrix is between 0.0070 and 0.010 cm. H is also known as height equivalent of a theoretical plate (HETP) and an equation for it is

$$H = \frac{L}{5.5}\left(\frac{W_{1/2}}{t_r}\right)^2 = \frac{L}{16}\left(\frac{W}{t_r}\right)^2 \tag{6.21}$$

Generally, Gaussian distribution is never observed and the peak appears asymmetric (Figure 6.19). The Dorsey –Foley equation for number of theoretical plates then is given as

$$N = \frac{41.7\left(t_r / W_{0.1}\right)^2}{a/b + 1.25} \tag{6.21}$$

where $W_{0.1}$ (the width of the peak at 10% of its height) $= a + b$.

The plate height can be minimized by resorting to different techniques and a few of them include: (i) reducing the particle diameter in the matrix, (ii) reducing the column diameter, (iii) changing column temperature, (iv) reducing the thickness of the liquid film, and (v) manipulating the flow rate of the mobile phase. Thickness of the liquid film depends on the physical properties of the continuous phase liquid (viscosity, density, surface tension, dielectric constant, etc.) and its interaction with the matrix.

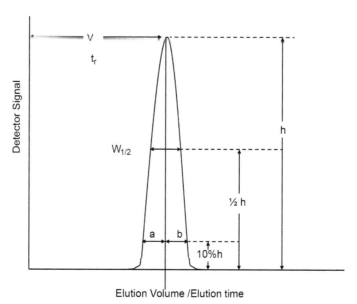

FIGURE 6.19   Chromatogram (detector signal).

Reducing particle or column diameter will lead to high back-pressure through the column.

Resolution between a pair of peaks i and j (R_{ij}) is defined as the distance of separation of the chromatographic peaks, i and j in a column. It is the distance of the peak maxima.

$$R_{ij} = \Delta t_{ij}/w_{av} = 0.589\ \Delta t_{ij}/W_{1/2av} \tag{6.21}$$

where Δt_{ij} is the separation between peaks and $W_{1/2av}$ is the average half width of the two peaks. If $\Delta t_{ij} = 2\sigma$, then $R_{ij} = 0.5$, If $\Delta t_{ij} = 3\sigma$, then $R_{ij} = 0.75$, If $\Delta t_{ij} = 4\sigma$, then $R_{ij} = 1.0$, If $\Delta t_{ij} = 6\sigma$, then $R_{ij} = 1.5$ (Figure 6.20).

Resolution between peaks improves with L, but it also leads to an increase in the elution time. Selectivity can be modified by changing the (i) composition of the mobile phase, (ii) changing the column temperature, (iii) changing the stationary phase, and (iv) using chemicals.

Capacity factor (k) of a compound indicates its retention behavior on a column.

$$k = \frac{K_d V_s}{V_m} = \frac{C_s/V_s}{C_m/V_m} \tag{6.22}$$

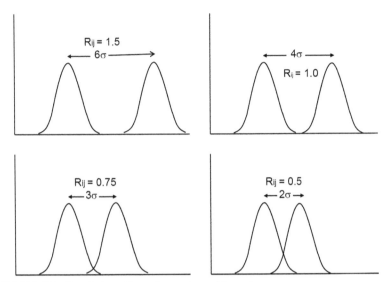

FIGURE 6.20   Resolution of two chromatographic peaks.

where K_d =distribution coefficient, V_s=volume of stationary phase, V_m =volume of mobile phase, C_s=solute concentration in the stationary phase, C_m= solute concentration in the mobile phase

6.1.11.1   Discrete Stage Analysis

If one assumes that the chromatographic column is not a continuous packed bed but a series of ideal stages or units as shown (Figure 6.21) and each unit is a well mixed tank of fixed volume in which the solution and adsorbent are in equilibrium. The stream flowing out is at the same concentration as it is inside the stage. Solvent flows from one stage to the next, but adsorbent remains within the same stage.

The solute mass balances in each stage,

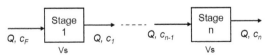

FIGURE 6.21   Multi-stage adsorption process.

$$\begin{bmatrix} accumulation \\ in\ liquid \end{bmatrix} + \begin{bmatrix} accumulation \\ in\ adsorbent \end{bmatrix} = \begin{bmatrix} solute \\ flow\ in \end{bmatrix} - \begin{bmatrix} solute \\ flow\ out \end{bmatrix}$$

$$\varepsilon V_s \frac{dc_n}{dt} + (1-\varepsilon)V_s \frac{dq_n}{dt} = Q(c_{n-1} - c_n) \tag{6.23}$$

where ε is the volume fraction of liquid (or void space) in the stage, V_s is the single stage volume ($=V/N$), c_n is the solute concentration in the liquid in stage n and also flowing out, q_n is the solute concentration in the adsorbent, and Q is the solvent flow (constant from one stage to another). V is the bed volume. N is the number of stages. C_{n-1} is the concentration entering stage n. This is called a difference differential equation and one could have n equations for n stages.

Solute concentrations in the solvent and in the adsorbent are assumed to be in equilibrium. If it is a linear isotherm relation then ($K =$ equilibrium constant)

$$q_n = Kc_n \tag{6.24}$$

Combining equations,

$$\left[(\varepsilon + (1-\varepsilon)K)V\right]\frac{dc_n}{dt} = Q\left(c_{n-1} - c_n\right) \tag{6.25}$$

Initially none of the stages contain solute, i.e., at $t< 0$, $c_n = 0$, $n =1, 2, \ldots, N$, where N is the total number of stages in the column. The solute is injected into the column (initial condition) at the first stage, $t = 0$, $c = c_F$.

The equations can be solved to arrive at a relation for concentration (c_n) of solute leaving stage n

$$c_n = c_F \left(\frac{\tau^{n-1} e^{-n}}{(n-1)!} \right) \tag{6.26}$$

where τ is given by

$$\tau = N \left\{ \frac{Qt}{\left[(\varepsilon + (1-\varepsilon))K\right]V} \right\} \tag{6.27}$$

When the number of stages is small, the concentration change as a function of time will be an exponentially decaying curve. As the number of stages becomes large, the profile approaches the Gaussian limit (Figure 6.22) whose equation is given below (this is similar to Eqs. (6.1) and (6.3))

$$c = c_0 \exp\left[-\frac{(\tau - \tau_0)^2}{2\tau_0^2 \sigma^2} \right] \tag{6.28}$$

In this Gaussian distribution, the concentration c is given as a function of three variables, τ_0, σ, and c_0. (Note: standard deviation equals $\tau_0 \sigma$). So,

$$\tau_0 = N \tag{6.29}$$

$$\sigma^2 = \frac{1}{\tau_0} = \frac{1}{N} \tag{6.30}$$

$$c_0 = \frac{c_F}{\sqrt{2\pi N}} \tag{6.31}$$

$$t_0 = \left[\varepsilon + (1+\varepsilon)K\right]V_B / H \tag{6.32}$$

$$c = \frac{c_F}{\sqrt{2\pi N}} \exp\left(-\frac{(t/t_0 - 1)^2}{2/N} \right) \tag{6.33}$$

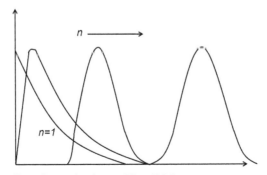

FIGURE 6.22   Effect of n on the shape of Eq. (6.26).

$$c = \frac{c_F}{\sqrt{2\pi N}} \exp\left(-\frac{(V/V_0 - 1)^2}{2/N}\right)$$

(6.33)

6.1.11.2 Continuous Packed Bed Model

If one considers a continuum packed bed (in the z axis) (Figure 6.23), then solute is (i) carried forward because of solvent flow, (ii) moves because of longitudinal diffusion, (iii) gets accumulated in the voids, and (iv) accumulates in the stationary matrix after adsorption or partition. This process could be represented as a partial differential equation in time and z axis as,

$$\varepsilon \frac{\partial c}{\partial t} + (1-\varepsilon)\frac{\partial q}{\partial t} = D_z \frac{\partial^2 c}{\partial z^2} - v \frac{\partial c}{\partial z}$$

(6.34)

where v is the superficial velocity, ε is the bed porosity. D_z is the diffusion coefficient along the bed axis. If one assumes that the diffusion $\left(D_z \frac{\partial^2 c}{\partial t^2}\right)$ is negligible and accumulation in the void space $\left(\frac{\partial c}{\partial t}\right)$ is also negligible then the equation simplifies to

$$(1-\varepsilon)\frac{\partial q}{\partial t} = -v \frac{\partial c}{\partial z}$$

(6.35)

The initial conditions are (i) there is no solute adsorbed on the packing before injection (i.e., all along z, $q = 0$ at time, $t<0$), and (ii) at $t = 0$, solute is just entering the bed.

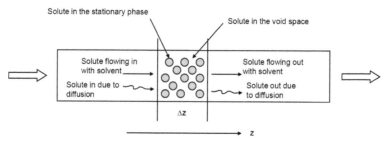

FIGURE 6.23   Continuous packed bed model.

For solving these equations one needs a relationship between q and c, which may depend on the rate controlling process. The controlling factors could be (i) mass transfer from the bulk of the solution to the surface of the particle or, (ii) diffusion inside the pores of the adsorbent.

If mass transfer of the solute from the bulk of the solution to the surface of the particle is controlling then

$$(1-\varepsilon)\frac{\partial q}{\partial t} = k_L a(c - c^*) \tag{6.36}$$

where k_L is the mass transfer coefficient, a is the packing area per bed volume, and c^* is the concentration of the solute in the solution which would be in equilibrium with the solute adsorbed on the adsorbent. If diffusion within the pores of the packing material is rate controlling, then the corresponding relationship has the form:

$$(1-\varepsilon)\frac{\partial q}{\partial t} = \sqrt{\frac{D}{\tau}}ac^* \tag{6.37}$$

where D is the effective diffusion coefficient in the pores and τ is a characteristic time.

Equation 6.35 in combination with one of the two equations needs to be solved to get a concentration profile as a function of Z and t. Each of these cases will give a different concentration profile.

If the chemical engineering concepts such as, number of transfer units (NTU) and height of transfer units (HTU) are brought into this analysis then, number of transfer units is equal to the number of stages or plates (N) and can be determined as described before. The length of the column will be

$$L = HTU.NTU \tag{6.38}$$

6.1.12 CHANGES IN THE STANDARD DEVIATION

The standard deviation also will be affected depending upon the mechanism as described above. If mass transfer of solute from liquid to solid is the controlling factor then,

TABLE 6.3   Effect of Various Parameters on k_L and σ

Controlling mechanism	k_L is proportional to
1. Diffusion of the solute from mobile phase (will be independent of the velocity, v, outside the particles)	$1/d$
2. Mass transfer between the bulk and the particle	$\left(\dfrac{v}{d}\right)^{1/2}$
Controlling mechanism	σ^2 is proportional to
1. Diffusion of solute within the particles is controlling (pore diffusion)	$\dfrac{v d^2}{l}$
2. If external mass transfer is the controlling	$\dfrac{v^{1/2} d^{3/2}}{l}$

$$\sigma^2 = \frac{v}{k_L a_L} \tag{6.39}$$

It is a function of the velocity, v, the column length, L, and the mass transfer coefficient, $k_L a$. The area per volume, a, for a packed bed of spherical particles is given by

$$a = \frac{f}{d}(1-\varepsilon) \tag{6.40}$$

where d is the diameter of the spherical particle, ε is the bed voidage, $f = 6$ for spherical particle and $= 4$ for long cylinders. The mass transfer coefficient, depending upon the controlling process, is proportional to various parameters as shown in Table 6.3.

So increasing velocity and particle size increases σ and hence, increases the width of the pulse. Increasing the column length decrease the width. The standard deviation may also change because of dispersion of flow in the column, due to polydisperse packing material.

6.2   DISTILLATION

Distillation, similar to liquid-liquid extraction, adsorption or gas-liquid absorption, is considered as an equilibrium staged process. Separation of liquids having different vapor pressures can be achieved by distillation.

The liquid with lower boiling point can be removed retaining the liquid with higher boiling point. If a liquid containing two components with different boiling points are made to reach an equilibrium with the vapor, the vapor will predominantly contain the component with low boiling point, and the liquid will contain predominantly the high boiling liquid (Perry and Green, 1997). The equilibrium concentration of the component with low boiling point in the vapor and in the liquid will generally appear convex as shown in the Figure 6.24. This is called a VLE (vapor–liquid equilibrium) diagram. The corresponding temperature of the vapor–liquid and mole fractions of the low boiling compound in the liquid and vapor will be as shown in Figure 6.25, and it is known as T-X-Y diagram. Using this diagram, one could determine the boiling point and composition of the vapor and liquid that are in equilibrium.

An ideal mixture is one which obeys Raoult's law and examples include hexane and heptane, benzene and methylbenzene and propan-1-ol and propan-2-ol. The Raoult's law states that the partial vapor pressure of a component in a mixture is equal to the vapor pressure of the pure component at that temperature multiplied by its mole fraction in the mixture.

For an ideal binary system

$$P_A = X_A P_A^{\,0} \tag{6.41}$$

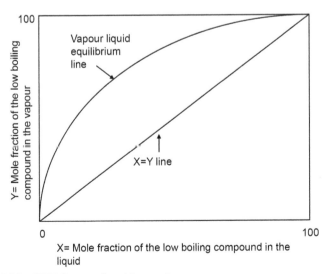

FIGURE 6.24   VLE diagram for a binary mixture.

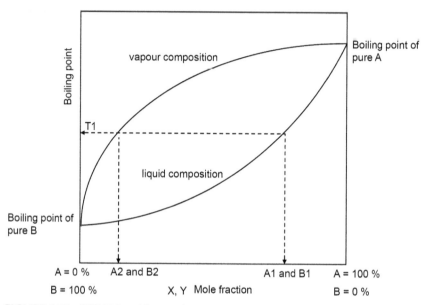

FIGURE 6.25   T-X-Y for a binary mixture.

and

$$p_B = X_B P_B^{\ 0} \tag{6.42}$$

where p_A and p_B partial vapor pressures of component A and B, respectively. X_A and X_B are the mole fractions of component A and B, respectively. $P_A^{\ 0}$ and $P_B^{\ 0}$ are the pure component vapor pressures of A and B, respectively.

The total vapor pressure of this mixture will be

$$p = p_A + p_B \tag{6.43}$$

$$\text{Also,} \ X_A + X_B = 1 \tag{6.44}$$

An azeotrope is a mixture of two or more liquids in such a ratio that its composition cannot be changed by simple distillation. When an azeotrope is boiled, the resulting vapor has the same ratio of the various components as it is in the liquid. Because the composition cannot be changed by distillation, azeotropes are also called constant boiling mixtures. A maximum-boiling azeotrope is one where the azeotrope temperature is higher than the boiling points of the individual components in the binary mixture (Figure 6.26a). A minimum-boiling azeotrope is one where the azeotrope temperature is less than the boiling points of the individual components in the binary mixture (Figure 6.26b). The VLE diagram for a binary azeotrope mixture may appear as shown in Figures 6.27a and 6.27b. Azeotrope can be broken by changing the operating pressure or adding an entrainer. The entrainer will carry one of the components with it either in the bottom or in the top of the distillation column. For example, a maximum boiling azeotrope such as, chloroform – ethyl acetate can be broken with an intermediate boiling entrainer, such as, 2-chlorobutane. A minimum-boiling azeotrope such as, acetonitrile-water mixture can be purified with hexylamine or butyl acetate. Pervaporation membrane process can also be used to break an azeotrope (Chapter 5, Section 5.4).

The distillation unit operation consumes high amount of energy, in terms of cooling and heating, and also contributes to plant operating cost.

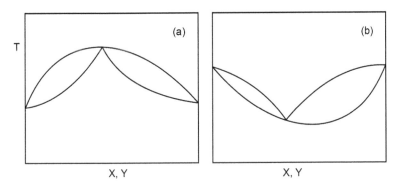

FIGURE 6.26   (a) High, and (b) low boiling binary azeotropes.

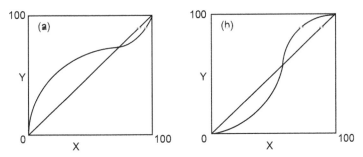

FIGURE 6.27   VLE diagram for a binary azeotrope.

The liquid mixture is heated in a rebolier at the bottom of the unit, which drives the vapor up the column. At the top it is cooled in a condenser. Part of the condensed liquid is sent or recycled back to the top of the column and this is called the reflux. The condensed liquid that is removed at the top is known as the distillate or top product. The reflux liquid flows down the column and encounters (or comes in contact) with the rising vapor. During this contact, the low boiling components in the liquid will join the rising vapor while the high boiling components in the vapor will join the liquid that is flowing down. This process makes the top product rich in the low density and the bottom product rich in the heavies. The fraction of liquid collected at the top that is refluxed back is called reflux ratio. Increasing reflux ratio improves separation efficiency, but increases distillation time as well as the energy cost. Changing the operating pressure can change the boiling points of the liquids and the separation efficiency.

Distillation can be performed in batch or continuous mode (Figures 6.28 and 6.29, respectively). Continuous distillation process can be classified as binary – where feed contains only two components or multi-component distillation – where feed contains more than two components. The number of product streams may be multi-product – where the column has more than two product streams.

Column may be tray (or plate) or packed and the internals provide better contact between the vapor and the liquid. In the former design, trays of various types are kept inside the column, to hold the liquid. Each tray

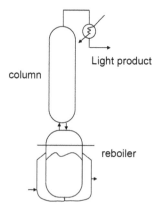

FIGURE 6.28 Batch distillation column.

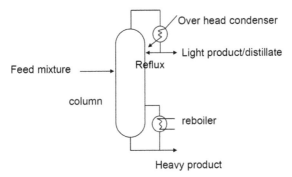

FIGURE 6.29   Continuous distillation column.

represents a distinct mixing place for the vapor and liquid. Under ideal conditions, the liquid flowing down and the gas going up from the tray will be in equilibrium. In the packed column, packing's of different designs are used and they are dumped (as random packing or structured packing). Packing provides not only enhanced contact area but also increases the turbulence.

Several mathematical correlations are reported in literature to determine the height of a distillation column or estimate the number of theoretical plates (or trays) in a column (Piché et al., 2003; Row and Lee, 1999). A book titled Distillation Design discusses various types of packing's, different types of distillation columns and design correlations (Kister, 1992).

6.3   CONCLUSIONS

This chapter deals with the product purification techniques especially, by chromatography and distillation. The former is predominantly used for purifying temperature sensitive biomolecules while the latter for small molecules. Chromatography uses packed beds of adsorbents like those used for adsorption. There are different types of chromatographic separation depending upon the physical principle. The separations are based on size, ionic nature, biological recognition, hydrophobicity, polarity, etc. The pulse is introduced in the entrance and it gets diluted with solvent and spreads as bell shape. The yield and purity of the product pulses are usually estimated by assuming that they have a Gaussian shape. This shape can be rationalized in terms of either equilibrium stages or transport processes. Several equations are available for determining the number of stages in a column. The number of stages in the column and the spread of the injected peak depend upon the controlling mechanism. Distillation is a high temperature process, which is suitable for thermally stable compounds. The theory and technology for distillation in industrial scale is well established over the past 100 years unlike chromatography.

KEYWORDS

- azeotrope
- chromatography
- column efficiency
- distillation
- Gaussian curve
- retention time
- thin layer chromatography
- vapor–liquid equilibrium

REFERENCES

1. Amersham Biosciences, Ion Exchange Chromatography, Principles and Methods, Amersham Pharmacia Biotech, Björkgatan 30, SE-751 84 Uppsala, Sweden, 2002.
2. Arnold, F. H., Blanch, H. W., Wilke, C. R., Analysis of Affinity Separations, Chem. Engr. J., 1985, 30, B25.
3. Ericksson, K.-O. Hydrophobic interaction Chromatography. *In* Protein Purification: Principles, High Resolution Methods and Applications. (J.-C. Janson and L. Rydén, eds.) VCH Publishers, New York. 1989, pp. 207–226.
4. Gibbs, S. J., Lightfoot, E. N., Scaling up gradient elution chromatography, IEC Fund., 1986, 25, 490.
5. Henry, Z. Kister, Distillation Design, McGraw-Hill Professional, 1992.
6. Janson, J. C., Hedman, P., Large scale chromatography of proteins, Adv. Biochem. Engr., 1982, 25, 43.

7. King, C. J., Separation Processes, 2nd ed., New York, McGraw-Hill, 1979.
8. Ladish, M. R., Volach, M., Jacobson, B., Bioseparations: Column design factors in liquid chromatography, Biotech. Bioengr. Sympos. Ser., 14, 525 (1984).
9. McAdams, W. H. Heat transmission, 3rd edn, McGraw Hill, USA, 1954.
10. Scott, R. P. W. *Techniques and Practices of Chromatography*; 2nd ed.; Marcel Dekker, 1995.
11. Yau, W. W., Kirlland, J. J., Bly, D. D., Modern Size-Exclusion Liquid Chromatography, Wiley, New york, 1979.

PROBLEMS

1. The retention time of a protein in a column is 9.0 min, and that the standard deviation is 0.4 min. Estimate the equivalent number of equilibrium stages in this column.

2. A protein is present as 70% and rest impurity in a mixture. Laboratory data suggest a retention time of 9.0 hr. with a standard deviation of 0.6 hr. for the former compound and 8. hr. with a standard deviation of 0.25 hr. for the latter. What will be the yield of the protein for a purity of 99%?

3. State the stationary and moving phases in each type of chromatography below:
   a) thin layer
   b) column
   c) gas

4. Two proteins A and B are chromatographed through a bed. The total volume of the bed is 100 L and void fraction is 0.30. The equilibrium constants K are 10.0 and 12.0 for A and B, respectively. Estimate the volume of eluate at which the peak will occur for each of these materials? When 600 and 700 L of eluate is collected the percentage of A was 2.0 and 16.0 respectively. Determine the quantity of eluate that must be collected to obtain a yield of 95% of A. Assume σ for the both solutes is the same.

5. What is the resolution of two Gaussian peaks of identical width (3.27 s) and height eluting at 67.3 s and 74.9 s, respectively?

6. About 15 g of a protein mixture is flowing through a column at a velocity of 30 cm/hr., the peak in concentration exists the column in 90 min and the standard deviation of this peak is given as 10 min. (a) How long must we purify for a 95% yield? (b) If we increase the flow to 50 cm/hr., how long must we run for this same yield if the process is controlled by diffusion through the packing? (c) How long must we wait if the process is controlled by mass transfer? (d) How long must we wait if the column actually contains equilibrium stages?

7. Calculate the number of theoretical plates when the time at which peak appears is 10 min and the width of the base is 1.2 min.

8. How many stages are required to recovery 90% of a protein if the flow rate is 10 L/min, porosity is 0.33, partition coefficient is 54, bed volume is 100 L.

9. Derive the mathematical equation relating the width at the base and width at half height in a chromatogram

10. Estimate the Resolution between two peaks A and B if the time at which peak maximum occurs are 8 and 12 min, respectively. The widths at the base are 3 and 4 min, respectively.

CHAPTER 7

PRODUCT POLISHING AND FINISHING

CONTENTS

This chapter deals with solid handling and its purification. If the product is a liquid, then these unit operations are not needed. In some cases the product (even a liquid) needs to be stabilized through suitable physico-chemical modifications for long-term storage to prevent it from oxidation or degradation.

7.1 CRYSTALLIZATION

One of the finishing steps in a product manufacture may involve crystallization to remove impurities, dry to remove any solvent/water present, prepare formulation to meet the requirements of the customer or achieve product stability. More than 80% of the substances used in pharmaceuticals, fine chemicals, agrochemicals, food and cosmetics are isolated or formulated in their solid form. Crystallization generally is the last chemical purification step and this unit operation can produce very pure product.

It is initiated either by cooling or by evaporation. The active ingredients of drugs are generally in the solid form. Transportation and storage of solid material is easier than a liquid product.

Both crystallization and precipitation involves separation of solids from a solution. Both are solid-liquid separation processes. In the case of the former solid crystals of defined shape and size is removed from a super saturated solution. The shape of the crystal can be cubic, tetragonal, orthorhombic, hexagonal, monoclinic, triclinic or trigonal. In order for crystallization to take place the solution must be "supersaturated" which is an equilibrium process. Supersaturation refers to a state in which the liquid contains more solute than can ordinarily be dissolved at that temperature.

The degree of supersaturation of a solution at a given temperature is measured as supersaturation coefficient

$$S = Ct/C0 \qquad (7.1)$$

where Ct = concentration of the solute in a solvent at a given temperature; $C0$ = concentration of the solute in the solvent in a saturated solution at the same temperature; $S = 1$ saturated solution; $S > 1$ solution is supersaturated.

$C0$ will depend on temperature. If the solution is supersaturated, spontaneous crystallization will occur until the solution becomes saturated. To maintain the crystallization process the liquid needs to be cooled further.

Nucleation is the first step in crystallization. This can occur due to the presence of impurities, on the walls of the vessel or can be initiated by adding seed crystals. Supersaturation can be achieved by altering the solubility of the solute in the solvent (or mother liquor) by (i) evaporating part of the solvent from the solution, (ii) decreasing the temperature, (iii) addition of a non-solvent which decreases the solubility of the solute. Addition of non-solvent may lead to impure crystals.

Usually as supersaturation is reached the instantaneous formation of many nuclei is observed and "crashing out" of the solution takes place. This is called the primary nucleation. In the next step of secondary nucleation, crystals start taking shape and then they grow in size. If the crashing out is not controlled properly, a large particle size distribution is obtained. Seeding is a better method to achieve uniform crystals. Each method of achieving supersaturation has its own benefits and disadvantages. For cooling and evaporative crystallization, energy expenditure is high. Large heat transfer surfaces are required for this operation.

Crystals are made of infinite number of unit cells. Unit cell is the smallest unit of a crystal, which if repeated, could generate the whole crystal. The unit cell dimensions are defined by six numbers, the lengths of the three axes, a, b, and c, and the three interaxial angles, α, β and γ. A cubic crystal is one where a = b = c and α = β = γ = 90°.

The rate of growth of a crystal depends on the transport of the material to the surface of the crystal and the mechanism of surface deposition. Stirring the solution helps in the transport of material to the surface. Rate of growth of crystals is a diffusion-controlled process. Presence of impurities reduces the rate of crystal growth.

Different types of crystallizer designs (Figures 7.1–7.4) are used and they are (1) batch tank crystallizers, (2) scraped surface crystallizers, (3) forced circulating liquid crystallizer (Figure 7.1), (4) evaporator-crystallizer (Figure 7.4), and (5) circulating magma vacuum crystallizer. The simple design is a batch vessel. Heat transfer coils and agitation are used to achieve the crystallization. Labor costs are high. This design is used only in fine chemical or pharmaceutical industries. In the second design, the agitator blade gently scrapes the vessel wall removing crystals that grow. The crystals do not get burnt since it is removed from the hot surface. The third design involves both evaporation and crystallization. The liquid is forced through the tube

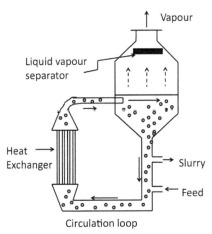

FIGURE 7.1   Forced circulation crystalliser.

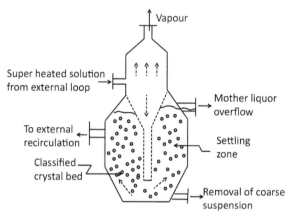

FIGURE 7.2   Classified suspension (Oslo type) Crystallizer.

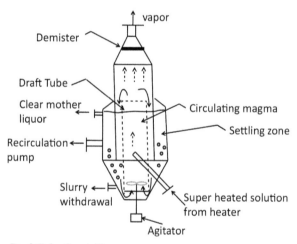

FIGURE 7.3   Draft Tube Crystallizer.

side of a heater (Figure 7.1). The heated liquid flows upwards into the vapor space of the crystallization vessel. Here, flashing of the solution takes place due to low pressure at the top of the vessel. This leads to a reduction in the amount of solvent in the solution (increasing solute concentration-leading to supersaturation). The supersaturated liquor flows down through a tube, producing nucleation and crystallization. Product crystals are withdrawn from the bottom and the liquor is recycled, mixed with the feed, and reheated. In the fifth design crystal/solution mixture (magma) is circulated, heated and mixed back into the vessel. A vacuum in the vessel causes boiling at the

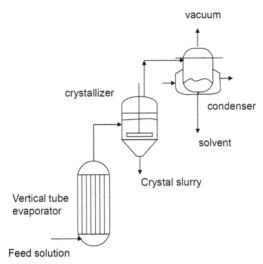

FIGURE 7.4 Evaporator cum crystallizer.

surface of the liquid (Figure 7.3). The evaporation causes crystallization and the crystals are drawn off near the bottom of the vessel body. Classified-suspension crystallizer is developed for the production of large, coarse crystals (Figure 7.2). The desupersaturation of the mother liquor is achieved by contact with the largest crystals present in the crystallization chamber and there is no stirrer provided here.

Crystallizer designs are based on empirical relationships and experience. A crystallizer is mathematically modeled as a two-phase system, one describing the solids (population balance model) and the other describing the supersaturated solution. The kinetics of micromixing can be represented as coalescence–dispersion of fluid element. The nucleation of the crystals will be dependent on the concentration of the solute in the solution (C) in excess of the supersaturation value (C_{su}). This difference acts as the driving force for the nucleation rate as shown below.

$$d[Nu]/dt = k \ (C\text{-}C_{su}) \qquad (7.2)$$

If the growth rate of crystals is dependent on diffusion, then the crystal mass (W) will depend on the mass transfer coefficient (k_{Sa}) and the concentration driving force ($C\text{-}C_{su}$)

$$dW/dt = k_{Sa} \ (C\text{-}C_{su}) \qquad (7.3)$$

Crystals of various sizes are formed which grow with time and hence, there will be a crystal size distribution similar to a Gaussian (Normal) distribution. It can be described as a population balance model. Some crystals will be older with larger size and some will be young, with smaller size. The dominant crystal size (d_c) will be

$$d_c = 3G\tau \qquad (7.4)$$

where G = crystal growth rate, τ = residence time in the vessel = V/F. V = volume of the vessel and F = feed rate. The fraction of crystals in this size range will be

$$= (1 - e^{-(dc/G\tau)}) \qquad (7.5)$$

McCabe law states that all crystals that are geometrically similar and have the same material in the same solution grow at the same rate. So, the growth is measured on the basis of increase in length and is independent of the initial size of the crystals

The overall transfer coefficient is same for each face of all crystals and so the growth rate will be

$$G = \Delta L / \Delta t \qquad (7.6)$$

The ΔL law is applicable for crystal growth of many materials at crystal sizes less than 0.3 mm but is not applicable to systems when crystals are subjected to different treatment procedures based on their size.

Recrystallization is done to recover as much of the solids present in the mother liquor. The crystals obtained during this recrystallization process may not be pure, but it may help to improve the yield of the product recovery. Hence, multiple stage crystallization is adopted (Figure 7.5). Crystallization is an energy intensive process. The vapor that is generated can be used to heat the next crystallizer to evaporate the liquid. This approach is known as multiple effect crystallizer (Figure 7.6). Such a design is common in sugar manufacturing industries.

The rate of heat required (Q) to evaporate water and the rate of evaporation of water (m) are related as

$$Q = m\lambda \qquad (7.7)$$

where λ is the heat of vaporization of water.

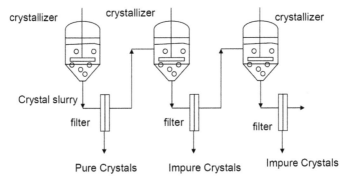

FIGURE 7.5   Multi-stage crystallizers.

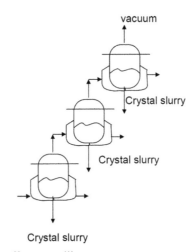

FIGURE 7.6   Multiple effect crystallizers.

A non-polar compound will dissolve well in a non-polar solvent, and a polar compound will dissolve well in a polar solvent. An ideal crystallization solvent should:

1. dissolve the entire compound at the boiling point of the solvent;
2. dissolve very little or none of the compound when the solvent is at room temperature;
3. have different solubilities for the compound and the impurities;
4. boiling point of the solvent must be below the melting point of the compound so that the compound dissolves, not melts, in the hot solvent;
5. have a relatively low boiling point;

6. will not react with the compound;
7. nonflammable;
8. nontoxic; and
9. cheap.

7.2 DRYING

Drying is an unit operation performed to remove water or solvent from the solid by heating and sometimes by applying vacuum so as to decrease the operating temperature. Drying is performed since a liquid product/formulation may be susceptible to chemical (such as, deamidation or oxidation) or physical (such as, aggregation and precipitation) degradation during storage. Other advantages of drying are convenience, especially drugs in tablet form; to reduce volume; may be necessary to remove undesirable volatile substances; more economical and convenient to store them in dry form rather than in frozen form; prevent bacterial growth, such as, sterilization and to recover expensive organic solvents. The solids obtained after drying could be the final product, intermediates or biomass. Maximum operating temperature in a dryer needs to be much below the permitted temperature of the solids that is being handled, in order to avoid product decomposition or charring.

During the drying process, the solids dry at two different rates. Initially, the water evaporates at a constant rate (Figure 7.7). This is called the constant rate period. When the moisture content in the water reaches a critical moisture value, the drying rate falls down. Here, moisture present in the

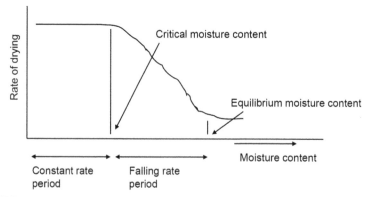

FIGURE 7.7   Drying process.

interstices or pores of the solid will diffuse out and gets evaporated. So the rate of drying keeps decreasing as the moisture content keeps falling. The moisture content in the solid reaches the equilibrium value and further drying stops. Equilibrium moisture content is a property of the solid. Drying takes place under constant and falling rate periods. In the former stage, the surface of the solid remains saturated with liquid water because of the movement of water within the solid to the surface is at a rate greater than the rate of evaporation from the surface. In the falling rate period, the rate of drying is controlled by the rate of movement of moisture within the solid and so the influence of air velocity on the drying process decreases. The amount of moisture removed in the constant rate period is relatively small, but the falling rate period is the major proportion of the overall drying time.

Drying of solids could be achieved through conduction, convection or radiation. Tray dryers or belt dryers where the support is heated on which the solids are placed are based on conduction (Figure 7.8a). The solids that are in contact with the bottom (tray, belt or pan) are heated by conduction. The governing equation in conduction is given as

$$Q = k \, A \, \Delta T / \Delta z \qquad (7.8)$$

where Q= rate of heat transfer, k = thermal conductivity of the material, A= area of contact, ΔT = is the temperature difference between the source and the solids and Δz = the distance.

Hot air ovens are based on convection where the hot air heats the solids (Figure 7.8b). The air is heated first and then made to flow over the material. The governing equation in convection is given as

$$Q = U \, A \, \Delta T \qquad (7.9)$$

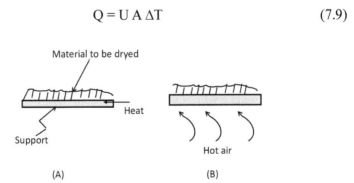

(A)   (B)

FIGURE 7.8   (A) Direct contact drying (B) Hot air drying.

where U is the overall heat transfer coefficient between the air and the solids. This depends on the physical properties of the solid and airflow rate. Several correlations have been reported in literature based on laboratory experimental data to estimate the heat transfer coefficient.

For fluids of viscosity close to water at turbulent flow (whether it is heating or cooling) (McAdams, 1954)

$$\frac{hd}{k} = Nu = 0.023 \, \text{Re}^{0.8} \, \text{Pr}^{0.4} \tag{7.10}$$

where Nu = Nusselt number; Pr = Prandtl number = $Cp\mu/k$; h = heat transfer coefficient; Cp, k = specific heat and thermal conductivity of the fluid; Re = Reynolds number = $du\rho/\mu$; ρ and μ are the density and viscosity of the fluid; u = velocity of the fluid; d = diameter.

Correlations for Convective Heat Transfer for many systems have been reported and they include forced convection flow inside a circular tube, forced convection turbulent flow inside concentric annular ducts, forced convection turbulent flow inside non-circular ducts, forced convection flow across single circular cylinders and tube bundles, forced convection flow over a flat plate, natural convection and film condensation, etc. (http://www.cheresources.com/convection.shtml, 10 Jan 2009).

The overall heat transfer coefficient (U) is made up of three transfer coefficients namely that of (i) the hot fluid side (h_h), (ii) the material separating them, and (iii) cold fluid side (h_c) as given below

$$\frac{1}{UA} = \frac{1}{A_h h_h} + \frac{1}{A_c h_c} + \frac{d}{kA} \tag{7.11}$$

where d = wall thickness, k = thermal conductivity of the material, A = overall contact area, A_h and A_c are the area on the hot and cold sides, respectively. If the wall thickness of a tube is small then all the areas are same. If plate type heat exchanger is used then the areas are the same. Thermal conductivity for some typical materials are, polypropylene = 0.12 W/mK; stainless steel = 21 W/mK and aluminum = 221 W/mK. Heat transfer coefficient for air = 10 to 100 W/ m²K and water = 500 to 10,000 W/ m²K.

Radiation could lead to high temperatures and is suitable for achieving very dry material. The heat transfer is proportional to surface reflectivity, emissivity, surface area, temperature, and geometric parameters. No medium needs to exist between the hot and cold bodies for heat transfer to take place.

The radiation energy per unit time from a blackbody is given by the Stefan-Boltzmann law as

$$q = \sigma\,A\,T^4 \tag{7.12}$$

where q = heat transfer per unit time (W); σ = 5.6703 10^{-8} (W/m^{2}K^4); T = absolute temperature Kelvin (K); A = area of the emitting body (m^2).

Different types of drying equipment include hot air dryer (Figure 7.8b), vacuum-shelf dryers (oven with trays) (Figure 7.9), rotary vacuum dryer, freeze dryers and spray dryers. Direct contact drying involves the material in direct contact with a heated surface (Figure 7.8a). Heat is supplied to the product mainly by conduction. Examples include drum dryer, roller driers and vacuum band driers. The necessary sensible heat and latent heat of evaporation are supplied to the material by conduction. Temperature sensitive food materials can be damaged in such dryers. If the drying is carried out under reduced pressure, lower surface temperature may be employed. Rotary vacuum dryer is a batch operation. Wet feed is charged and is subjected to indirect heating while undergoing agitation due to the action of a paddle. The operation is carried out in vacuum. Recovery of solvent is possible here by condensing the vapor generated.

Spray dryer produces a dry powder from a liquid or slurry by rapidly drying with a hot gas (Figure 7.10). It is preferred for thermally sensitive

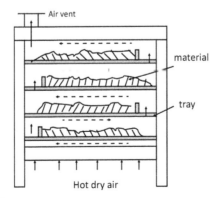

FIGURE 7.9   Tray dryer.

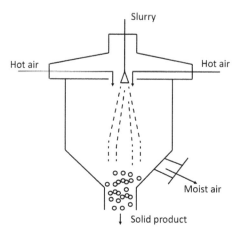

FIGURE 7.10   Schematic of a spray dryer.

materials such as, foods and pharmaceuticals. Hot air is the drying media. Nitrogen is used for oxygen-sensitive or flammable solvent such as, ethanol. It is widely used in food (milk powder, coffee, tea, eggs, cereal, spices, flavorings) and pharmaceutical (antibiotics, medical ingredients, additives) industries. The important design parameters in a spray dryer are: (i) evaporation rate, and (ii) particle size distribution of the product. The former determines the amount of air needed for drying, which in turn determines the size and cost of the equipment. The latter affects the choice of atomization and hence, determines the size of the spray dryer. Very short drying time (1–10 sec) and low product temperature can be achieved in this dryer. The droplets of the spray usually are of 10–200 µm in diameter, leading to large surface area per unit volume of the material.

In freeze dryers, removal of water from the material is achieved after it is initially frozen. This is suitable for heat labile materials including proteins, antibiotic, vitamins, blood plasma, hormones, tissue and micro-organisms.

7.3   LYOPHILIZATION

Biological materials often must be dried to stabilize them during storage. Lyophilization is achieved by freezing the wet substance and then causing the ice to sublime directly to vapor by flashing it to low pressure (Figure 7.11). The latter converts the ice directly into water vapor. The

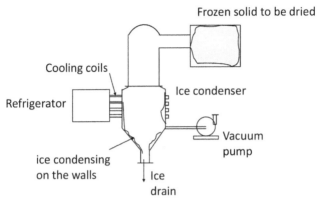

FIGURE 7.11   Schematic of a freeze dryer.

primary mechanism here is sublimation. The various steps in freeze-drying are:

1. Cooling of the product to a sufficiently low temperature to allow complete solidification.
2. Reducing the pressure in the chamber to below the vapor pressure at the triple point of water so that drying can occur by sublimation. Ice, water and water vapor exist in equilibrium at the triple point (0.01°C and 4.6 mm Hg). Below this triple point, water passes directly from the solid to the vapor phase by sublimation when the temperature is raised.
3. Unbound water is removed in a drying phase called primary drying.
4. A higher shelf temperature and additional time are required to remove the bound water in the secondary drying phase. The time to complete the drying cycle is ~ 24 to 48 h.

Because the liquid phase is not present after the product is initially frozen, reactions that occur in the liquid phase such as, hydrolysis, cross-linking, oxidation, aggregation, and disulfide rearrangement are eliminated. Larger the surface area of the frozen material, the faster is the rate of lyophilization and thicker is the frozen material, slower is the rate of lyophilization.

The simplest lyophilizer consists of a vacuum chamber with facility to freeze the contents. It is operated in batch mode, which consists of freezing at atmospheric pressure followed by applying a low pressure to remove the ice into water vapor. Another design consists of having two chambers one for freezing and the other is operated at low pressure. Of course, many micro-organisms and proteins survive lyophilization. It is useful to dry vaccines, pharmaceutical drugs, blood fractions, diagnostics and specialist food products by this method. The energy and capital costs of lyophilization are 2–3 times higher than other drying methods. But for majority of high-value proteins, peptides, and vaccines, lyophilization is the only way to prepare stable, biologically active products with long shelf-life.

7.4   PRODUCT STABILIZATION

Biological products are produced by fermentation, biotransformation, separation and purification and they are generally unstable due to the aqueous environment. For long-term preservation of the activity, the amount of water in the product must be reduced. Freezing or drying are a few methods to maintain its shelf-life. In addition, stabilizing agents such as, antioxidants, radical scavengers, preservatives, or antimicrobial natural products are added. These additives should be removable, if needed, later.

Enzymes need to be immobilized to achieve long-term stability (Figure 7.12). Native enzymes are labile and may get denatured very quickly. Techniques for immobilization are broadly classified as (i) entrapment, (ii) covalent binding, (iii) cross-linking, and (iv) adsorption. Entrapment is the best technique for the immobilization of cells, but not for enzymes. Possible slow leakage during continuous use is a disadvantage in this technique. Biocatalysts have been entrapped in natural polymers including agar, agarose and gelatine through polymerization. In alginate and carrageenan, ionotropic gelation technique is adopted. A number of synthetic polymers

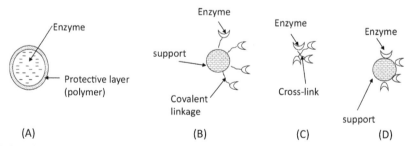

FIGURE 7.12 Techniques for the immobilisation of enzymes (A) entrapment (B) covalent binding to a support (C) cross-linking (D) adsorption to a support.

including photo-cross-linkable resins, polyurethane and acrylic polymers including polyacrylamide have also been found effective. Glutaraldehyde has been used to cross-link enzymes.

A medium consisting of a polyhydroxy compound (5–60% by weight of the medium) and phosphate ions (molar ratio of phosphate ions to hydroxy groups in the polyhydroxy compound is = 0.025–0.625) has been tested to stabilize and store biological compounds. (US Patent 6653062 – Preservation and storage medium for biological materials, Issued on November 25, 2003).

Proteins in solution rapidly deteriorate. In spite of purification very small traces of impurities can slowly initiate degradation or deterioration processes. For a proteinaceous preparation, inactivation could be retarded by (i) addition of chemical additives [such as, glycerol, $(NH_4)_2SO_4$]; (ii) undercooling (subzero temperatures, unfrozen); (iii) chemical modification/immobilization; (iv) sequence alteration/"protein engineering"; (v) freeze/thaw, lyophilization; or (vi) addition of water-soluble glasses (Franks, 1993).

KEYWORDS

- **crystallization**
- **drying**
- **heat transfer**
- **lyophilization**
- **product stabilization**
- **supersaturation**

REFERENCES

1. Arakawa, et al., Stabilization of Protein Structure by Sugars, Biochemistry, 21, 6536–6544, 1982.
2. Brown, Theodore L., Chemistry: The Central Science, 5th Ed., Prentice Hall, New Jersey, 1991, ISBN: 0-13-126202-5.
3. Geankoplis, Christie J., Transport Processes and Unit Operations, 3rd Ed., Prentice Hall, New Jersey, 1993, ISBN: 0-13-930439-8.

4. http://www.cheresources.com/convection.shtml, Online Chemical Engineering Information, 10 Jan 2009.
5. McAdams, W. H. Heat Transmission, 3rd edn., McGraw Hill, USA, 1954.
6. Price, Chris J., "Take Some Solid Steps to Improve Crystallization," *Chemical Engineering Progress*, September 1997, p. 34.
7. Protein Biotechnology Isolation, Characterization, and Stabilization, The Humana Press Inc. 1993, DOI 10.1007/978-1-59259-438-2_14, Felix Franks, Chapter 14. Storage Stabilization of Proteins, pp. 489–531.
8. Swenson Process Equipment. Website at www.swensontechnology.com
9. US Patent 6653062 – Preservation and Storage Medium for Biological Materials, Issued on November 25, 2003.

PROBLEMS

1. What is the heat required to evaporate 1000 kgs of water at 100°C?
2. The length of a crystal increased from 0.1 mm to 0.15 mm in 10 min. What is the growth rate if the crystal growth follows McCabe's law?
3. If a copper cylinder of length 5 cm and diameter of 2 cm is heated from one end at 100°C, while the other end is at 30°C, what is the rate of heat transfer. If the material is made of iron what will be the heat transfer rate?
4. Air at 70°C is heating solids at 30°C by convection. The area of heat transfer is 10 cm^2 and the heat transfer coefficient is 2 kJ/cm^2/K/s. What is the rate of heat transfer? What are the techniques one can adopt to improve the rate of heat transfer?
5. For fluid similar to water, if the velocity is doubled what will be the increase in the heat transfer coefficient? If the viscosity is doubled what will be the change in the heat transfer coefficient?
6. In radiation heat transfer, if the temperature of the black body increases from 250 to 260°C, what will be the increase in the radiation energy?
7. What are the different techniques that can be adopted to double the rate of drying in conduction?
8. In spray drying if the particle size is reduced from 120 microns to 100 microns, what will be the increase in heat transfer rate?
9. If the heat transfer coefficient of the hot and the cold fluid sides are neglected, what will be the change in the overall heat transfer coefficient if aluminum is used instead of stainless steel?
10. If 120 kcals/sec of heat is applied to water for evaporation, what will be the rate of evaporation of water?

CHAPTER 8

UTILITIES AND AUXILIARY PROCESSES

CONTENTS

This chapter deals with utilities used in a biochemical process industry such as, air, water, steam, coolant and other auxiliary process such as, sterilization. Air is required in aerobic fermentor. Water is used in the process as well as a cooling the liquid in reactors, condensers and crystallizers. Coolant is required for condensing gases or cooling the contents of a liquid. Steam finds application in distillation columns, evaporators, etc., for heating the contents. Usage of utilities determines the product variable cost or product operating cost. Higher is this value, higher will be the product variable cost. Hence, minimum usage of utilities is always desired.

8.1 HEATING

Steam will be required for sterilization, heating the reactor contents or in distillation, evaporation, drying or crystallization operation. In an evaporator a heat balance between the heat taken up by the process fluid from steam can determine the amount of steam required for the process. Q_s is the amount of steam required to heat a process fluid so that mL amount of process fluid gets heated up from a temperature of T_i to T_0 and m_s amount of process fluid gets evaporated. The energy balance equation will be as shown as equation 8.1. The second term in the left hand side of the equation indicates the sensible heat lost by the condensed steam when it condenses to a temperature of T_w from T_s (Figure 8.1). λ_s is the heat of vaporization of the process fluid (fluid that is evaporated) and λ is the corresponding value for water.

$$Q_s \lambda - Q_s C_p (T_s - T_w) = mL C_p (T_i - T_0) + ms \lambda_s \qquad (8.1)$$

This equation does not assume heat loss to the ambient. That will be an extra term in the right hand side.

8.2 STERILIZATION

To prevent bacterial growth, the media and all other raw materials required for the bioprocess need to be sterilized. Two commonly adopted approaches are thermal and microwave. The problems with the first approach are deactivation of proteins and enzymes, food, syrups, etc., while better control

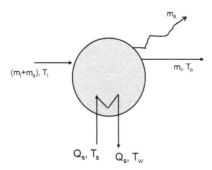

FIGURE 8.1 Energy balance.

can be achieved with the second method. But local high temperature spots are a problem here. Rapid heating and rapid cooling is a good approach to prevent deactivation of temperature sensitive materials as shown in Figure 8.2

The mode of heating could be Batch or Continuous. In the batch process, the contents are taken in a stirred vessel, heated and then discharged. While in the second approach, the fluid continuously enters a tube from one end and leaves from the other end (Figure 8.2). The first approach can lead to over heating.

8.3 MIXING

The mixing of liquids, solids and gases is one of the commonest unit operations in chemical, pharmaceutical and food processing industries. It is carried out to bring about intimate contact between various phases and prepare new formulations or emulsions. Mixing is done either by mechanical or non-mechanical means and its performance is judged by the degree of mixedness (uniformity) achieved, time required for mixing and power requirement.

8.3.1 LIQUID MIXING (AGITATION)

Agitation is employed for mixing gas to liquid, liquid-to-liquid, solid to liquid and solid-to-solid. This is done in order to suspend solids in liquid to form a slurry; blending miscible liquids (alcohol and water); dispersing a gas through the liquid especially, in fermenters; dispersing a second liquid

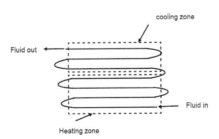

FIGURE 8.2   Continuous sterilizer (rapid heating and cooling system).

that is immiscible with the first, to form an emulsion or fine drops (oil in water); enhancing heat transfer between the liquid and a heating coil or jacket; and maintaining uniform temperature, concentration and pH in a vessel.

Different types of impellers/agitators/stirrers are used (Figures 8.3(a)–(e) and (g)) and they impart different flow patterns to the liquid. Agitator is selected based on the rheology of the liquid, desired shear rate, size of the vessel, nature of operation and mixing pattern. Chapter 10 also discusses the concept of agitation and mass transfer. Mixing vessel is rounded at the bottom to prevent the formation of dead spaces. Baffles (generally four in number) are vertical metal plates running the full depth of the inside surface of the tank.

8.3.2   SOLIDS MIXING (BLENDING)

Mixing of solids is carried out in paint and food processing industries (Figures 8.3 (f) and (h)). The solids mixing takes place due to diffusion, convection, and shear. Particles diffuse due to concentration gradient in

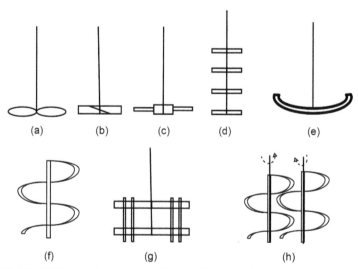

FIGURE 8.3   Different stirrer designs (a) propeller (b) pitched blade paddle (c) turbine–flat blade (d) multiple paddles (e) anchor (f) screw (g) gate (h) kneader.

the same way as molecules diffuse in liquid. Fick's law of diffusion can describe this phenomenon. Solids also move in void space.

Segregation of particles takes place during motion due to differences in their physical properties (density, porosity, shape, size, etc.). For example in pharmaceutical industry, the size and densities of the various solid components used in the manufacture of pills and tablets are almost equal. Electrostatic forces and surface changes also lead to segregation. Soft particles may form a coating on hard ones due to the mechanical forces. Mixing of solids are carried out in ribbon blender (Figure 8.4), kneader (Figure 8.5), cone mixer (Figure 8.6), tumbling mixer (V or Y type) (Figure 8.7), Kneaders, namely dough and paste mixers, are used for mixing solids and are operated at high power which is dissipated in the form of heat, and

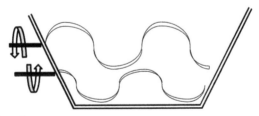

FIGURE 8.4   Ribbon blender.

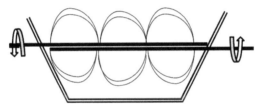

FIGURE 8.5   Kneader.

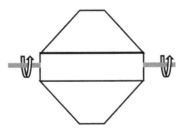

FIGURE 8.6   Cone mixer.

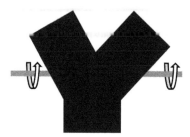

FIGURE 8.7   Tumbling mixer.

may spoil the product. Jacketing with cooling water circulation is needed to remove heat.

The ribbon blender consists of a long and horizontal trough in which two helical screws rotate in opposite direction. It is good for thin pastes or slurries. Tumbling mixers (Figure 8.7)/cone mixers (Figure 8.6) are two conical containers which rotate horizontally. The cones could be V, Y or cube shaped. They are operated in batch mode, filled with 60% by volume solids and run at 100 rpm. They are used for free-flowing solids and suitable for particles of similar size and density because strong segregation can occur here. Baffles are fitted inside which help to lift the solids and plows fitted inside assist convection.

8.4   BOILERS

Boilers produce (low, medium or high pressure) steam, which are used in evaporators, fermenters and distillation columns. They are classified according to the pressure, materials of construction, size of the tubes, contents (for example, waterside or fireside), firing, or circulation. Boilers are also classified by their heat source, such as, oil-fired, gas-fired, coal-fired, or solid fuel-fired boilers. Fire tube boilers consist of tubes that are housed inside a water-filled outer shell. Hot combustion gases flow through the tubes, which heat the water. Oil, coal or natural gas is used to produce the combustion gases.

In water tube boilers, hot combustion gases are circulated outside, while tubes are filled with water. An upper steam drum at the top separates the steam from water. They are more efficient and can produce higher pressure steam them the fire tube boilers. Fire tube boilers are cheaper, higher fuel efficiency and easy to operate. Electric boilers use electric heating coils immersed in water and are of very low-capacity units. Tubeless boilers are like jacketed pressure vessels with water located between the shells. Combustion gases are fired inside the inner vessel, which heats the water.

8.5   HEAT EXCHANGERS

A heat exchanger is an equipment where a fluid is heated or cooled. The transfer of thermal energy is between two or more fluids or between a solid surface and a fluid. Typical applications involve heating or cooling of a fluid stream, evaporation or condensation of fluid streams, recover or reject heat, sterilize, pasteurize, distil, concentrate or crystallize a process fluid. Common examples of heat exchangers include shell-and tube type exchangers (Figure 8.8), automobile radiators, condensers, evaporators, air preheaters, and cooling towers. If no phase change occurs in any of the fluids in the exchanger, it is called a sensible heat exchanger. Mechanical devices are used to enhance heat transfer as in scraped surface exchangers, agitated vessels, and stirred tank reactors.

To increase the heat transfer area, fins are connected to the surface. According to the way they are constructed, heat exchangers are clarified as tubular, plate type, extended surface (fin) and regenerative. Based on the process function, they are classified as condensers, vaporizers, heaters,

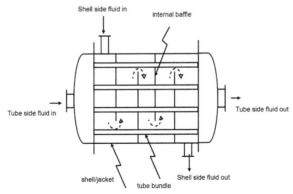

FIGURE 8.8 Single pass shell and tube heat exchanger or (1,1) heat exchanger. (Movement of fluid in the shell side is shown in dotted arrows).

coolers or chillers. Based on the flow arrangements they are clarified as single pass (Figure 8.8), counter flow, parallel flow and cross-flow or multi-pass (Figure 8.9) shell and tube or plate and frame type. There could be multiple shell or multiple tubes. Direct contact heat exchangers may have immiscible fluids, gas-liquid or liquid-vapor systems. Indirect contact systems may have a heat storage vessel as an intermediate. Heat exchangers are constructed with metals, alloys, or non-metals depending on the fluid that is handled.

Shell and tube heat exchangers consists of multiple tubes through which one fluid flows, while the other fluid flow in the shell side the fluid flowing through the shell and tube side could be made to pass more than once through the unit (Figures 8.8 and 8.9).

Plate heat exchanger (Figure 8.10) consists of thin plates joined together, with a small space between them. The surface area is very large, and hot and cold fluids flow in alternate plates. They and are more efficient than shell and tube heat exchangers.

8.6 COMPRESSORS AND COMPRESSED AIR SYSTEMS

Compressed air (Figure 8.11) is used in process applications, aerobic fermenters, for transfer of material from one vessel to another, pneumatic

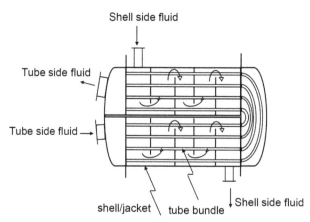

FIGURE 8.9   Multi-pass shell and tube heat exchanger or (2,1) heat exchanger. (Tube side two passes and shell side one pass. Movement of fluid in the shell side is shown in dotted arrows).

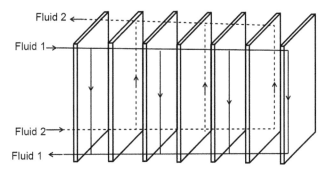

FIGURE 8.10   Plate heat exchanger.

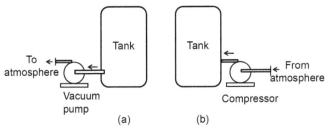

FIGURE 8.11   Arrangement of (a) vacuum pump and (b) compressor.

conveying, or operates instruments. Size of compressed air units range from 5 to over 50,000 hp. The operating cost of a compressed air system is higher than the cost of the compressor itself. A typical system consists of a intake air filter, inter-stage cooler, after-cooler, air-dryer, moisture drain trap and receiver.

Two basic types of compressors are positive-displacement (or recip-rocating) and dynamic. In the former, air is trapped in a chamber and is compressed to a high pressure and discharged. Fresh air is again drawn in, and this operation is repeated. So the output will be pulsating. In a dynamic compressor energy is imparted to air by means of impellers rotat-ing at very high speed.

The reciprocating air compressor is considered single or double acting when the compressing is accomplished using only one side or both sides of the piston. The entire compression can be carried out in a single cylin-der (single stage) or many cylinders in parallel (multiple stage). The air is normally cooled between stages.

Rotary compressors have rotors, which give a continuous discharge free from pulses. They operate at high speed and give higher through-put than reciprocating compressors. They are cheaper and compact than the former. Types of rotary compressors include: lobe compressor, screw compressor or rotary vane/sliding-vane, liquid-ring, and scroll-type.

In centrifugal air compressor energy is transferred from a rotating impeller to the air. This is a continuous compressor and suitable for high volume applications.

Free air discharge from a compressor (under isothermal conditions in Nm^3/s) is

$$Q = \frac{P_2 - P_1}{P_0} \frac{V}{t} \tag{8.2}$$

where, P_2 = final pressure after filling (kg/cm^2 a); P_1 = initial pressure (kg/cm^2 a) after bleeding; P_0 = atmospheric pressure (kg/cm^2 a); V = storage volume in m^3 which includes receiver after cooler and delivery piping; t = time taken to build up pressure to P_2 in sec.

If T_2°C is the discharge temperature and T_1°C is the ambient air temperature, then the free air discharge is to be corrected by a factor $(273 + T_1)/(273 + T_2)$.

Compressor efficiency can be calculated as volumetric efficiency, isothermal efficiency and specific power consumption as given below:

$$\text{Isothermal Efficiency} = \frac{\text{Actual Measured Input Power}}{\text{/Isothermal Power}} \tag{8.3}$$

$$\text{Isothermal Power (kW)} = P_1 \times Q_1 \times \ln r/36.7 \tag{8.4}$$

where, P_1 = absolute intake pressure kg/cm^2; Q_1 = free air delivered m^3/hr.; r = pressure ratio P_2/P_1.

$$\text{Volumetric Efficiency} = \frac{\text{Free Air Delivered (in } m^3/min)}{\text{/Compressor Displacement}} \tag{8.5}$$

where, compressor displacement $= A D^2/4 L S n$; D = Cylinder bore, m; L = Cylinder stroke, m; S = Compressor speed rpm; A = 1 for single acting; A = 2 for double acting cylinders; n = No. of cylinders.

The specific power consumption is the most appropriate way of comparing the performance of different compressors and it is defined as kW power/volume flow rate.

8.7 VACUUM PUMPS

These pumps generate vacuum (Figure 8.12) and they are opposite of compressors. Here, the discharge rather than the intake, is at atmospheric pressure. Vacuum maybe required in reactors to remove vapors or gases produced. It may be required in boilers and distillation columns to decrease the boiling point and for transfer of liquids from one vessel to another without resorting to the use of a pump.

Vacuum pumps (similar to air compressors) are either positive displacement type or non-positive displacement machines. The former draws constant volume of air. Different types of positive displacement vacuum pumps are piston, diaphragm, rocking piston, rotary vane, lobed rotor,

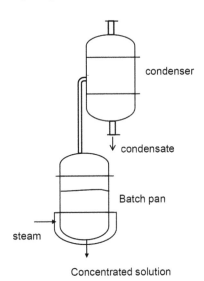

FIGURE 8.12 Batch evaporator.

and rotary screw designs. Non-positive displacement vacuum pumps use changes in the kinetic energy to remove the air from a system. They include centrifugal, axial-flow, and regenerative designs.

The free air that needs to be removed is = volume of the vessel/reactor (increased by 15–20% to allow for leakages) × desired vacuum in atmosphere.

Volumetric efficiency changes with vacuum level multiplied by vacuum level in atmosphere.

$$\text{Time required to pump the vessel from } p_1 \text{ to } p_2 = \frac{V}{S}\ln\frac{p_2}{p_1}$$

where v = system volume; p_1 and p_2 = initial and final pressures, respectively, in absolute units; S = pump capacity in cu.ft./min at the actual pressure in the system (average between p_1 and p_2).

8.8   EVAPORATORS

Concentration of a product present in a solution can be accomplished by boiling out the solvent or water in an evaporator. It is an energy intensive process. The water or solvent vapor is collected in a condenser once again as liquid. It is a unit operation used extensively in foods, chemicals, pharmaceuticals, fruit juices, dairy products, paper and pulp industries. Different types of evaporators are used in industries, which include:

- batch pan;
- forced circulation;
- natural circulation;
- wiped film;
- rising film tubular design;
- falling film tubular design; and
- rising/falling film tubular.

Batch pan are commonly used in salt industries, effluent treatment and food industries (Figure 8.12). They are of simple design and have residence time in hours or days. The batch pan is jacketed or has internal coils or heaters. It is not suitable for heat sensitive or thermo-degradable products. Heat transfer areas are normally small so they have agitation within the vessel to improve it. Fouling of the heat transfer surface is a problem here.

Tubular evaporators have internal tubes. The circulation of the vapor and liquid could be forced (that is with the help of a pump) or natural. In the former the boiling is prevented within the unit by virtue of a hydrostatic head. The liquid flashes to form a vapor at the separator located at the top. In natural circulation system there is a circulatory pump. The liquid and vapor rises through this exchanger tube (short) creating a natural circulation due to density differences. The vapor separates at the top of the vessel and leaves while the liquid comes down.

Rising film tubular evaporators consist of vertical tubes with steam condensing on the outside surface and liquid inside the tube, which is brought to a boil (Figure 8.13). The vapor generated forms a core in the center of the tube and the liquid touches the tube in the form of a thin film. Higher vapor velocity is achieved here that result in very high heat transfer

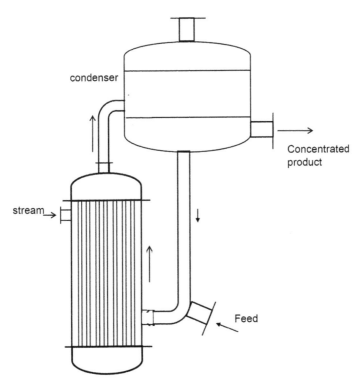

FIGURE 8.13   Natural circulation evaporator.

coefficient and shorter residence time. The product is removed from the top of the evaporator.

In falling film tubular evaporators the liquid is distributed along the walls of the tubes, which descends down due to gravity. This results in a thinner, faster moving film, short product contact time and good heat transfer coefficient. The rising film evaporator requires a driving film force, a large temperature difference than a falling film evaporator.

The wiped film evaporators are good for very viscous and concentrated materials and the stripping of solvents to very low levels is achieved. Feed is introduced at the top of the tubes and is spread by wiper blades. Evaporation of the solvent takes place as the thin film moves down the wall. A disadvantage of this evaporator is the need for moving parts. This system suitable for syrups, fruit juices, etc.

KEYWORDS

- blending
- boiler
- compressor
- evaporator
- heat exchanger
- mixing
- sterilization
- vacuum pump

REFERENCES

1. Bloch, H. P., Hoefner, J. J. (1996). Reciprocating Compressors, Operation and Maintenance. Gulf Professional Publishing. ISBN 0-88415-525-0.
2. Coulson, J., Richardson, J. (1983), Chemical Engineering – Design (SI Units), Volume 6, Pergamon Press, Oxford.

3. Felder, R. M., Rousseau, R. W. Elementary Principles of Chemical Processes (2nd Edition), John Wiley, 1986.
4. Foust, A. S. et al., Principles of Unit Operations (2nd Edition), John Wiley, 1980, pp. 500–501.
5. McCabe, W. L., Smith, J. C., Harriott, P. Unit Operations of Chemical Engineering (5th Edition), McGraw-Hill, 1993.
6. Perry, Robert H., Green, Don W. (1984). Perry's Chemical Engineers' Handbook (6th ed.). McGraw-Hill. ISBN 0-07-049479-7.
7. Saunders, E. A. (1988). Heat Exchanges: Selection, Design and Construction. New York: Longman Scientific and Technical.

PROBLEMS

1. How long it will take to reduce the pressure of a 100 L tank from 1 Kg/cm^2 (a) to 0.2 Kg/cm^2 (a), if the pump capacity is 500 L/min. assume 10% leakage.
2. In a rising film evaporator, if the heat transfer coefficient is proportional $V^{0.6}$, where V is the rising vapor velocity. If the throughput through the reactor has to be increased by 25%, how many more extra tubes of 25 cm^2 internal diameter are required?
3. What will be change in the heat transfer coefficient in problem 2 after the modification?
4. Calculate the average force air discharge from a compressor to compress a tank of 1000 L 1 kg/cm^2 to 5 kg/cm^2 at 30°C 2 h.
5. Calculate the isothermal efficiency of compressor discussed in problem 4.
6. Calculate the volumetric efficiency discussed in problem 4, if it contain 4 single action cylinders.
7. Calculate the energy required to remove 10% of water in a batch pan containing 10,000 L of water at 100°C.

CHAPTER 9

FUTURE TRENDS

CONTENTS

The downstream purification steps in chemical processes are well-established and therefore, newer trends and innovations will be limited and far between. Improvement in packing design used in absorption columns, design of novel trays in distillation columns, and newer gas-liquid contacting equipment are a few research areas, which will be pursued.

Downstream process is a mandatory requirement to manufacture any biopharmaceutical drug or biologics. The extent of purification depends on the type of product and the cost of the product in turn depends on the complexity of purification. As mentioned before the selling price of a product depends on the number of downstream steps, more so with bioproducts. The downstream process in bioindustry market is US$1.5 bn in the year 2009, and expected to reach US$ 3.75 bn in 2016 (GBI Research, 2012). The total reagent market is US $4.301 bn in 2009, and expected to grow to US$6.640 bn. These numbers indicate the size of this business and the importance the industries place to downstream processing.

Eric S. Langer, of BioPharma Associates in his article "Tracking Downstream Purification-Downstream is viewed as a capacity constraint now, and figures to remain that way" (Ref: http://www.pharmamanufacturing.com/articles/2009/059.html, Apr 2011) indicates that 48% of USA companies and 68% of West European companies (in 2009) felt Downstream processing in biotechnology is the capacity constraint which means although they have sufficient capability to expand based on their fermentor/bioreactor sizes they are not able to do so due to the size limitations in the downstream units. The top concerns of the Bioindustry (2000) are cost of chromatography steps for purification and how to move away from Protein A as an affinity chromatography ligand. Industry survey has identified depth filtration (32% of the respondents), chromatography column (43%) and ultrafiltration (26%) as the three purification steps that create capacity constraints in bioindustries (Masser, 2014). Protein A is a 40–60 kDa MSCRAMM (microbial surface components recognizing adhesive matrix molecules) surface protein originally found in the cell wall of *Staphylococcus aureus*. It is encoded by the *spa* gene and its regulation is controlled by DNA topology and cellular osmolarity. Adhesin proteins mediate the initial attachment of bacteria to host tissue, leading to infection. Protein A binds with high affinity to human Immunoglobulin G (*IgG*) and IgG2, and mouse IgG2a and IgG2b. Protein A is immobilized on a solid support and used for purifying total IgG from crude protein mixtures including serum, or coupled with one of the above markers to detect the presence of antibodies.

When industries personnel were asked to list the downstream steps that will cause bottleneck (McGlaughlin, 2012) in the future they come up with the following (starting from the most serious):

1. column chromatography;
2. virus removal;
3. ultrafiltration;
4. depth filtration;
5. tangential flow filtration;
6. clarification;
7. sterile filtration;
8. diafiltration.

Nearly two-thirds of biopharmaceutical companies are spending for acquisition of new downstream processing technologies in 2010. It is also believed that improvements are expected in the following areas:

*   membrane processes;
*   monoclonal-fragment technology;
*   specialty resins and synthetic affinity matrices that cost as close to as Protein A.

Examples of a few downstream technologies that is being patented by Bioprocess are listed below

- Novozymes's Disposable Dual Affinity Polypeptide technology replacing Protein A process (Novozyme, 2002).
- Design of stimuli responsive polymers, which enable complexation and manipulation of proteins and allow for control of polymer and protein complex solubility by Millipore Corporation. This helps in the direct capture of the product without centrifugation or Protein A media.
- Mixed mode sorbents to replace Protein A and ion exchange, with improved selectivity, capacity with and short residence times by Pall Corporation. These include hydrophobic charge induction chromatography, with low/no salt.
- Monoliths columns containing chromatography medium as a single-piece from BIA Separations (Strancar et al., 2002).
- Simulated moving beds, from Tarpon Biosystems which consists of multicolumn countercurrent chromatography system (Bisschops et al., 2009).
- Camel-derived antibodies to IgG from BAC.
- New inorganic ligands, including synthetic dyes, from Prometic Biosciences.
- Expanded bed adsorption chromatography systems, from Upfront Chromatography.
- Ultra-durable zirconia oxide-bound affinity ligand chromatography media from Nysa Membrane Technologies.
- Membrane affinity purification system from Pure Pharm Technologies.
- Peptidic ligands for affinity chromatography from Prometic Biosciences and Dyax.
- Protein A- and G-coated magnetic beads from Invitrogen and Dynal.
- New affinity purification method based on expression of proteins or MAbs as fusion proteins with removable affinity tag for licensed by Roche (Genentech).
- Plug-and-play units with disposable components, design of experiments capability for process development, and multicolumn continuous capture, from GE Healthcare.

According to Robert Blanck, bioprocess strategic marketing manager at Millipore Corporation (Yavorsky et al., 2003) the future direction in downstream process is integrated disposable process solutions, which include bioreactor, chromatography, TFF, virus filtration, sterile filtration, buffer/media storage, and intermediate handling.

Future discoveries or changes in downstream will be in the following areas:

- affinity chromatography;
- alternates to Protein A (Including reverse micelles (liposomes), liquid–liquid extraction systems, crystallization, immobilized metal affinity chromatography, and membrane chromatography systems);
- new packing materials in column chromatography;
- virus filtration;
- cross-flow membrane filtration;
- simulated moving bed;
- single use disposables units;
- plug and play units;
- mobile and modular biopharma operation in separation and chromatographic steps (GBI research, 2012).

It is very clear from the reports from the Bioindustry leaders that major innovations have to be carried out in downstream for the industries to be competitive since they are the bottle-neck rather than the reaction.

KEYWORDS

- **affinity purification**
- **column chromatography**
- **disposable reactor**
- **plug-and-play systems**
- **protein A**
- **virus filtration**

REFERENCES

1. Ales Strancar, Ales Podgornik, Milos Barut, Roman Necina, Short Monolithic Columns as Stationary Phases for Biochromatography Advances in Biochemical Engineering/Biotechnology Volume 76, 2002, pp. 49–85.

2. David Yavorsky, Robert Blanck, Charles Lambalot, and Roger Brunkow, The Clarification of Bioreactor Cell Cultures for Biopharmaceuticals, Pharmaceutical Technology, March 2003, 62–76.

3. Derek Masser, Enabling Greater Process Control and Higher Protein Titers: Advances in Downstream Single-Use Technologies, Bioprocess International, June 1, 2014 (http://www.bioprocessintl.com/downstream-processing/separation-purification/enabling-greater-process-control-and-higher-protein-titers/).

4. GBI Research, Downstream Processing in Biopharmaceuticals – Adoption of Disposable Technology at Improved Economies of Scale to Optimize Production Efficiency and Cost-effectiveness, November 1, 2012 (http://www.marketresearch.com/GBI-Research-v3759/Downstream-Processing-Biopharmaceuticals-Adoption-Disposable-7183194/).

5. Langer, E. S. http://www.pharmamanufacturing.com/articles/2009/059.html, Apr 2011.

6. Marc Bisschops, Lynne Frick, Scott Fulton, Tom Ransohoff, Single use, continuous countercurrent multicolumn chromatography, Bioprocess International, 7(6), (June 2009).

7. Molly S. McGlaughlin, An Emerging Answer to the Downstream Bottleneck, Bioprocess International, May 1, 2012 (http://www.bioprocessintl.com/analytical/downstream-development/an-emerging-answer-to-the-downstream-bottleneck-330225/).

8. Novozyme Patent, Dual affinity polypeptides for purification, WO 2009062942 A2, Filing date Nov 12, 2008.

CHAPTER 10

FUNDAMENTAL CONCEPTS

CONTENTS

10.1 INTRODUCTION

This chapter briefly lists the important concepts of mass, heat and momentum transfer. Since it is not within the scope of this book to dwell in detail about the basic principles of chemical and biochemical engineering, some additional reference books are listed in the reference section at the end of this chapter, which discusses these aspects in detail.

10.2 MASS AND ENERGY BALANCE

The basis of design of a downstream processing unit is mass and heat balances. In an open system for any species,

Material In = Material Out +/– Material Produced Or Consumed.

Mass balance is very important to know the efficiency of any operation and waste produced

Similarly, for heat transfer the heat balance will be,

Heat Input + Heat Generated = Heat Output
+ Heat Loss To Surroundings.

Heat loss includes heat lost to the ambient or heat removed by the cooling system. Section 10.4 also discusses the concepts of mass and heat balances.

10.3   MIXING

The goal of mixing in a vessel is to achieve uniform concentration of the various species (including gases). The speed with which complete mixing is achieved is important. If the reaction is fast and the mixing is slow this will be the rate controlling step. If the mixing is slow, one need to understand where the bottle-neck is. Mechanically agitated stirrers are use in chemical reactors, fermentors, extractors, etc. to achieve good mixing. Chapter 8 sketches various types of agitators to achieve good mixing. Section 1.6 discusses the rpm required for solid suspension and good heat transfer. Non-mechanical methods of mixing include sparging gases through a liquid bed which would create a circulation pattern or allow the fluids to pass through a stationary packed bed which would allow them to come in intimate contact and mix.

The uniformity of mixing in agitated vessels depends on the pumping action brought about by the agitator. If there are no baffles in a vessel mechanical agitation will bring about a vortex (Figure 10.1a) and the entire liquid will move (swirl) without any mixing. Baffles are long, flat plates that are attached to the side of the tank to prevent vortex formation and also promote top to bottom fluid movement. Baffles increase the energy required for mixing. With different types of agitators radial or axial mixing could be achieved (Figures 10.1b and 10.1c, respectively). Different types of agitators require different amounts of power to achieve the same mixing velocities. Initially, power number decrease as the Reynolds number is increased in a linear fashion (Figure 10.2). For very high Reynolds number the power number remains constant.

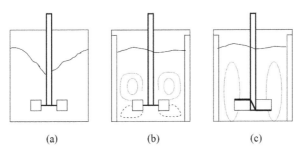

FIGURE 10.1 Various types of mixing patterns in a cylindrical vessel with a mechanical agitator (a) absence of baffle leading to vortex (b) radial mixing (c) axial mixing.

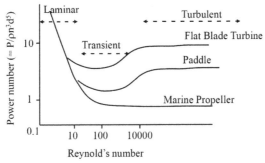

FIGURE 10.2 Power number vs. Reynolds number for agitated vessels and different agitator geometries (P = power input, r = density of the fluid, n = rotational speed and d = diameter of the impeller).

10.4   MASS TRANSFER AND GAS LIQUID CONTACTING

In aerobic fermentation, gas-solid or gas-liquid biochemical or chemical reactions, oxygen gas has to diffuse from the gas phase into the bulk of the solution and then reach the bacterial cell, cell cluster, solid catalyst or solid reactant (depending upon the case may be). This transport has to overcome several resistances.

For example, in a gas liquid transfer, resistance is offered by gas side film as well as the liquid side film (Figure 10.3). One of the films may be controlling depending upon the type of gas liquid contacting equipment used. The thickness of these films depends upon on several factors including physical properties of the fluid and intensity of mixing. In an aerobic fermentation process, the movement of oxygen gas from the bulk gas phase (i.e., air) first diffuses through the stagnant film in the liquid side

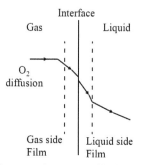

FIGURE 10.3   Movement of oxygen from gas to liquid phase.

surrounding the gas bubble, diffuses through the bulk liquid, then diffuses through the stagnant film surrounding the bacterial single cell or a floc and finally reach a single cell (Figure 10.4).

Mechanical and non-mechanical gas liquid contacting approaches are shown in Figure 10.5. They are used in environmental applications as well in chemical and biochemical reactors and absorbers, etc. Mechanical agitation also helps in breaking the bubbles so that the surface area per unit volume increases.

For a binary mixture of A and B, the molar flux, J_A, resulting from the concentration gradient along the x axis is

$$J_A = -cD_{AB} \, dy_A/dx \qquad (10.1)$$

where, c is the total molar concentration, D_{AB} is the diffusivity of A in B, y_A is the mole fraction of A. For a binary mixture $D_{AB} = D_{BA}$.

The total molar flux, N_A, due to diffusion and bulk flow then becomes

$$N_A = J_A + y_A(N_A + N_B) \qquad (10.2)$$

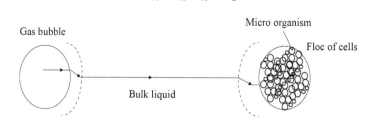

FIGURE 10.4   Movement of oxygen during an aerobic biochemical process and the various resistances.

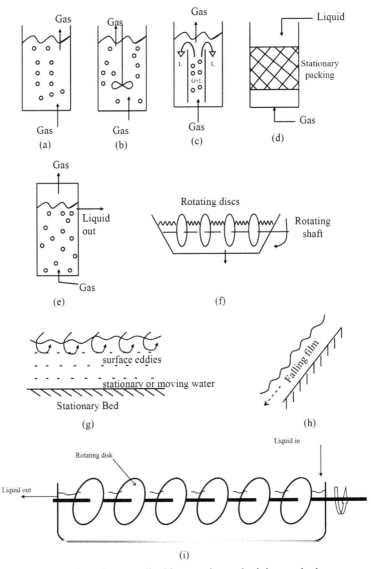

FIGURE 10.5   (a)–(i) Various gas–liquid contacting and mixing methods.

N_B is the molar flux of B. The term $y_A(N_A + N_B)$ is the molar flux resulting from the bulk flow of the fluid.

If mass flows from the bulk to the inside of a bubble or a catalyst, then the mass transfer equation will be

$$\text{Flux at the interface} = k_L \, a \, (c_L - c) \qquad (10.3)$$

C_L and C are the concentrations at the bulk and at the interface respectively. Several mathematical correlations are reported in the literature which can be used to calculate the mass transfer coefficient (k_L) as well as a, depending upon the type of fluids and mixing patterns.

10.5   HEAT TRANSFER AND EQUIPMENT

Similar to mass transfer the equation for heat transfer will be,

$$\text{Heat flux} = U \times A \times \Delta T \qquad (10.4)$$

where U is the overall heat transfer coefficient, A is the contact area and ΔT is the temperature driving force. The overall heat transfer coefficient is made up of several terms, namely hot fluid side film heat transfer coefficient (h_c), cold fluid side film heat transfer coefficient (h_h) and resistance offered by the material of construction that is separating the two fluids (Figure 10.6) as,

$$1/U = 1/h_c + 1/h_h + t/(A \times k) \qquad (10.5)$$

where, t is the wall thickness, A is the area and k is the thermal conductivity of the material of construction. Several mathematical correlations are

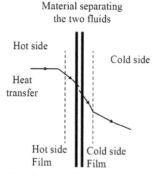

FIGURE 10.6   Flow of heat from hot to cold side.

reported in the literature, which can be used to calculate the heat transfer coefficient depending upon the type of fluids and the mixing patterns.

The various types of heat transfer such as, conduction and convection specifically in drying are described in Chapter 7. Heat transfer equipment include heat exchangers, evaporators, dryers, condensers, and boilers, etc. Fouling of heat transfer surfaces will reduce the effectiveness of the heat exchangers.

10.6 MOMENTUM TRANSFER AND FLUID FLOW EQUIPMENT

In a closed system (no exchange of matter with the outside takes place and is not acted on by outside forces) the total momentum is constant. This law of conservation of momentum is implied by Newton's laws of motion. Three fundamental physical principles upon which all of fluid dynamics is based on are, mass is conserved; Newton's second law holds good; and energy is conserved.

Fluid flow equipment include pumps, compressors, blowers and vacuum pumps.

KEYWORDS

- fluid flow
- gas liquid contacting
- heat transfer
- mass and energy balance
- mass transfer coefficient
- mixing

REFERENCES

1. Coulson, J. M., Richardson, J. F., Coulson and Richardson's Chemical Engineering: Fluid Flow, Heat Transfer, and Mass Transfer (6th Ed.), Butterworth-Heinemann, 1999.

2. Himmelblau, D. M., Riggs, J. B. Basic Principles and Calculations in Chemical Engineering, 7th Ed., Prentice hall International series, May 2012.
3. Jay Bailey, James Bailey, David F. Ollis, Biochemical Engineering Fundamentals Hardcover, Feb, 1986.
4. Kern, D. K., Process Heat Transfer, Tata McGraw-Hill Education, India, 1950.
5. Michael L. Shuler, Fikret Kargi, Bioprocess Engineering: Basic Concepts (2nd Edition), Pearson Education Ltd., Nov, 2001.
6. Warren McCabe, Julian Smith, Peter Harriott, Unit Operations of Chemical Engineering, 7th Ed., McGraw-Hill, 2004.

INDEX

9 781774 635544